Social Origins and Primal Law

Andrew Lang
& J. J. Atkinson

Social Origins and Primal Law

The present edition is a reproduction of previous publication of this classic work. Minor typographical errors may have been corrected without note, however, for an authentic reading experience the spelling, punctuation, and capitalization have been retained from the original text.

ISBN: 978-1-64799-645-1

TO

ANNABELLA ALLEYNE

Dear Annie,

As you first pointed out to me the facts which are the germ of my Theory of the Origin of Totemism, you are one cause of my share in this book. The other is affection for the memory of the author of 'Primal Law.'

Yours always,
A. LANG

St. Andrews:
Feb. 13, 1903

INTRODUCTION

The portion of this book called 'Primal Law' is the work of the late Mr. James Jasper Atkinson. Born in India, of Scottish parents (his mother being the paternal aunt of the present editor), Mr. Atkinson was educated (1857-1861) at Loretto School, then managed by Messrs. Langhome. While still young he settled on certain stations in New Caledonia bequeathed to him by his father, and, except for visits to Australia and a visit to England, he lived and died in the French colony. His ingenious mind was much exercised by the singular laws and customs of the natives of the New Caledonian Archipelago and the adjacent isles. These peoples have been little studied by competent European observers—that is, in New Caledonia. Mr. Atkinson wrote an account of native manners before he had any acquaintance with the works of modern anthropologists, such as Mr. Tylor, Mr. McLennan, Lord Avebury, and others. To these he later turned his attention; he joined the Anthropological Institute, and, in the course of study and observation, he discovered what he conceived to be the 'Primal Law' and origin of morality, as regards the family. In his last illness, in 1899, he was most kindly attended by Commander John Haggard, R.N., then Her Majesty's Consul in New Caledonia. Mr. Atkinson's mind, in his latest moments, was occupied by his anthropological speculations, and, through Mr. Haggard, he sent his MS. to his cousin and present editor. I have given to it the last cares which the author himself would have given had he lived. But I have also taken the opportunity to review, in the following pages, introductory to 'Primal Law,' the present state of the discussion as to the beginnings of the rules regulating marriage among savages.

The discussion is now nearly forty years old, if we date it from the appearance of Mr. J. F. McLennan's Primitive Marriage in 1865. Yet, in spite of the speculations of some and the explorations of other distinguished students, the main problems are still in dispute. Was marriage originally non-existent? Was promiscuity at first the rule, and, if so, what were the origins, motives, and methods of the most archaic prohibitions on primitive license? Did man live in 'hordes,' and did he bisect each 'horde' into exogamous and intermarrying moieties, and, if he did, what was his motive? Are the groups and kindreds commonly styled 'totemic' earlier or later than the division into a pair of moieties or 'phratries'? Do the totem-kins represent the results of an early form of exogamous custom, or are they additions to or consciously arranged subdivisions of the two exogamous moieties? Is a past of 'group marriage' or 'communal marriage' proved by the terms for human relationships employed by many backward races, and by survivals in manner and custom?

These are among the questions examined in the introductory chapters that may be read either before or after Mr. Atkinson's Primal Law. To him I am indebted for the conception of sexual jealousy as a powerful element in the evolution of exogamy.

Since my attention was first directed to these topics, I have felt that a clear and consistent working hypothesis of the origin of totemism was indispensable, and such an hypothesis, with a criticism of other extant theories, is here offered. Throughout I have attempted to elucidate and bring into uniformity the perplexing and confused special terms employed in the discussion. Here it should be explained that by 'marriage' in this work I mean permanent cohabitation of man and woman, sanctioned by tribal custom, and usually preceded by some rite or initiation which does not prelude to casual amours. By family or fire circle I mean the partners to this permanent cohabitation, their offspring, and such kinsfolk by blood or affinity as may be members of their camp. In the first sentence of the book I speak of the family as 'most ancient and most sacred,' and I do so deliberately. The primitive association described I take, with Mr. Darwin and Mr. Atkinson, to be 'most ancient,' and to be the germ of the historic family, which is 'most sacred.' But to 'sacred' when I apply the word to the primitive fire-circle I give no religious sense, such as the Greek hearth enjoyed under Hestia, youngest and oldest daughter of Zeus. I mean that the rules given to the primitive fire-circle by the sire were probably the earliest and the most stringent, though not yet sanctioned by a tabu or a goddess.

Such a small circle, and not a promiscuous horde or commune, I conceive, with Mr. Darwin and Mr. Atkinson, to have been the earliest form of human society.

The book deals only with the institutions of races certainly totemistic, and mainly with the Australian and North American tribes, which present totemism in the most archaic of its surviving forms. But little is said, and that tentatively, on the question as to whether or not the ancestors of the great civilised peoples, ancient and modern, have passed through the stage of totemic exogamy, as our evidence is weak and disputable. Too late for citation in the body of the book I read Mr. A. H. Keane's theory of the origin of totemism.[1]

Mr. Keane's theory is much akin to my own as it stood in

---

[1] Man, Past and Present, Cambridge, 1899, pp. 396, 397.

Custom and Myth (1884) and to that of Garcilasso de la Vega, the oldest of all. Garcilasso (1540-1616), an Inca on the mother's side, describing the animal and plant worship of the low races in the Inca Empire, says 'they only thought of making one differ from another and each from all.'[2] But it may be that he had not totemism in his mind; the passage is not too explicit.

Mr. Keane says: 'And thus the family, the initial unit, segments into a number of clans, each distinguished by its totem, its name, its heraldic badge—which badge, becoming more and more venerated from age to age, acquires inherited privileges, becomes the object of endless superstitious practices, and is ultimately almost deified.... Its origin lies behind all strictly religious notions, and it was at first a mere device for distinguishing one individual from another, one family or clan group from another.[3] Thus among the Piaroas of the Orinoco below San Fernando de Atabapo the belief holds that the tapir, originally the totem of the clan, has become their ancestor, and that after death the spirit of every Piaroa passes into a tapir; hence they never hunt or eat this animal, and they also think all the surrounding tribes are in the same way each provided with their special animal fore-father. It is easy to see how such ideas tend to cluster round the clan[4] or family totem, at first a distinguishing badge, later a protecting or tutelar deity of Protean form. It should be remembered that the personal or family name precedes the totem, which grows out of it, as seen by the conditions still prevailing amongst the very lowest peoples (Fuegians, Papuans of Torres Strait[5]).'

I am indebted in various ways to assistance, chiefly in the interchange of ideas, from Mr. A. C. Haddon, Mr. G. L. Gomme, Miss Burne, and Mr. A. E. Crawley, author of The Mystic Rose. Mr. Crawley kindly read the book, or most of it, before publication, and collaborated most efficiently in the way of suggesting objections. It is not implied that any of these students accept the ideas of the two authors. I regret that it has been found impossible to wait for the publication of a new book by Mr. A. W. Howitt, from which we may expect much new information.

The question of the relations of religion and totemism is scarcely touched on in this work. A certain amount of regard is given to their totem animals and plants by some of the Australian tribes, to the extent of not killing, plucking, or eating them, except

---

[2] Royal Commentaries, i. 47.
[3] The Import of the Totem, Amer. Ass., Detroit, 1897.
[4] M. Chaffanjon, Tour du Monde, 1888, lvi. 348.
[5] Ethnology, pp. 9, 11.

under stress of need, but even this is not universal. There also exists, in some cases, a sense of kinship with them. They are not worshipped. That magic is worked for their preservation and propagation, as by the Arunta, proves nothing in the nature of a religious attitude towards them. In my opinion this religious regard for the totem does not appear till ancestor worship, which does not occur in Australia, has made considerable advance and a myth arises that an ancestral spirit or family god is incarnate in the animal which originally was only a totem. If so, totemism is not an element in the origins of religion, but a field later invaded by religion.

On the other hand, Dr. Achelis, of Bremen, writes that to savage man 'animals are his equals. To the ancient worship of animals is added, under the influence of sympathetic emotion, the worship of ancestors and totemism, which sees in a beast worshipped as a god the ancestor of the whole tribe.'[6] Clearly this sentence is replete with errors and confusions. The whole tribe, in Australia, does not regard any animal as its ancestor. No beast is worshipped as a god. No ancestors are worshipped. If the animals are 'his equals,' why did man worship them, and that apparently before the worship of ancestors and totemism arose? In an essay like that of Dr. Achelis on Ethnology and Religion the facts ought to be correctly ascertained.

I have been obliged to place in Appendix A certain facts about group names derived from animals which came late to hand, among them Mr. Robertson's interesting letter on many such names in the Orkneys, and some remarks on village names derived from animals among the ancient Hebrews.

---

[6] The International Quarterly, Dec.-March, 1902-1903, p. 321.

CONTENTS

SOCIAL ORIGINS

CHAPTER I

THE EARLY HISTORY OF THE FAMILY

THE FAMILY. THEORY OF MR. ATKINSON

The Family is the most ancient and the most sacred of human institutions; the least likely to be overthrown by revolutionary attacks. In epochs of change the Family naturally invites the attentions of impetuous reformers, like Shelley (who advocated a scheme more than any other apt to shock the conscience of a savage), and like the friends of 'Free Love,' who would introduce a license beyond the Urabunna model. The horror aroused by certain relations, such as that of brother-and-sister marriage, is perhaps the oldest of moral sentiments, yet it has lost its hold of some barbaric races, and has been overcome by dynastic pride, as in the Royal House of the Incas of Peru, and in that of Egypt. While the Family, everywhere almost, has been secured by a religious and all but instinctive dread of certain aberrations, the laws or customs which may not be broken have varied in different lands, and in different stages of civilisation. What is incest in one age or country is innocent in another; still certain unions, varying in various regions, have always been regarded with loathing. No such emotion is known to be felt among the lower animals, and scientific curiosity has long been busy with the question, why should the least civilised of human races possess the widest list of prohibited degrees? What is the origin of the stringent laws that, among naked and far from dainty nomads, compel men and women to seek their mates outside of certain large groups of real or imagined kindred? The answers given to this question have varied with the facts of savage law which chanced to be at each moment accessible to inquirers, and all attempts to solve the problem must be provisional. New knowledge may upset even the most recent theory, and, indeed, new knowledge of the rules of certain Australian tribes has already produced fresh hypotheses, as regards certain aspects of the problem.

The whole subject is thorny, and I must crave pardon for venturing to differ, provisionally, on several important points, from authorities whose learning, research, and experience far exceed my own. The facts which they have collected from personal knowledge of savages, and from reading, often group themselves otherwise in my eyes than in theirs—the perspective is different. My observations, therefore, are submitted to criticism with all diffidence. Only the main lines of a complex discussion are here traversed, and the works cited are, as a rule, either by English-speaking authors, or, at least, are sometimes accessible in English translations. It will be seen that students have differed greatly, not only from each other, but, at different times, from themselves, under the influence of new facts brought in from the most remote and isolated of savage races. One author is most interested in this, another in that, factor of the problem. The difficulty of the subject cannot be exaggerated; for the origins of our human society cannot be historically traced behind the institutions of the races now lowest in the scale of culture. We are driven to risk hypotheses. Again, it is by no means certain that some of these lowest peoples of to-day (say the Arunta of Central Australia) represent a moment in the main current of the stream of tendency, a point through which all progress has passed. The ideas and institutions of such tribes may be mere local 'sports,' other divergencies may have arisen in other quarters, and it would be an error (repudiated by Mr. McLennan, the founder of the study in England) to suppose that, everywhere, exactly the same series of changes evolved itself in due sequence. 'In one place or another everything may have been going on,' I have heard Mr. McLennan observe.

Once more, the subject is obscure because the races apparently 'nearest the beginning,' the naked Australians, houseless hunters, just emerging from the palæolithic condition as regards implements, are, as to society and system of thought, very far from being 'primitive;' very remote from 'the beginning.' Their social rules are various and extremely complex, especially as regards marriage: some of their social customs are perhaps inexplicable—a field for modern guesswork—their speculative philosophy is, in one instance, ingenious, elaborate, and highly peculiar. The 'beginning' lies far behind them, yet their society and institutions may have their germs (on the Darwinian theory) in a state of all but complete brutality.

To trace human institutions back to that hypothetical stage of first emergence from the brute is the purpose of the following treatise, 'Primal Law,' by Mr. Atkinson. It were superfluous for me to dwell on the audacity of his enterprise. Of thoroughly human man we know a good deal: of the brutes we know something. Of a hypothetical creature, not wholly brute, but not yet 'articulate-

2

speaking man,' we know nothing, and as to the ways of his supposed next of kin, 'the great extant anthropoid apes,' our knowledge is vague, resting on the accounts of native observers. Such a creature, however, half ape, half human, is in part the theme of Mr. Atkinson's speculations, on which I venture to express no opinion: as not being persuaded that man ever had such a direct ancestor.

PRIMITIVENESS IN MAN

As to men really primitive, and their social arrangements, I only venture to conjecture that, in the nature of the case, they probably lived a nomadic life, 'selecting a temporary place of abode, whether a cave, rock, shelter, or hut, influenced chiefly by the amount of edible materials to be found in the neighbourhood.'[7] The area of the wandering of each group of hearth-mates would be limited, probably, by the existence of other groups, which would resent poaching. A large trout may often be seen to turn angrily and drive away a little trout that has ventured too near the bend of the brook which the large trout finds a good station for flies; and human groups would also, as in cases to be cited they do, mortally resent intrusions. I conceive that the males would be polygamous (like the gorilla) and jealous, killing or expelling the young males, as in the theories of Mr. Darwin and Mr. Atkinson. Thus groups would, on the whole, be hostile,[8] 'wandering from one locality to another, now gathering fruits and seeds, now hunting wild animals, or, as a last resource, feeding on shell-fish and other produce of the shore.'[9] The implements now used by backward savages for fish-catching, nets, spears, and barbed hooks, cannot be precisely primitive. Primitiveness, we must remember, does not depend on antiquity of date.

The Australians, though now their groups have coalesced into local tribes in defined areas, and though their customary law is extremely complex, are least remote from the primitive, least remote, but very far removed. They are, though our contemporaries, infinitely beneath the status in culture of palæolithic man of the mammoth and reindeer period. It is not improbable that he had domesticated the ox, goat, pig, horse, and dog. 'They manufactured fine needles of bone, with which they sewed their skin garments.

---

[7] Dr. Munro, Archæological Journal, vol. lix. no. 234, pp. 109-143: (Tire à part, p. 1.) See also later, Hypothetical Early Groups.
[8] To this point, hostility, I return later.
[9] Dr. Munro, Archæological Journal, vol. lix. no. 234.

They adorned their persons with a variety of beads....' Their art was of notorious and amazing excellence. Dr. Munro says that they were 'ignorant of the rearing of domestic animals,'[10] but also that 'there seems to be no inherent improbability in the idea that some of them' (ox, goat, horse, pig, and dog) 'had been domesticated by the indigenous inhabitants prior to the coming of the neolithic brachycephals into France.'[11] A palæolithic sketch of a horse 'with a supposed cover,' and another of a horse with a bridle,[12] may be misinterpreted: Dr. Munro thinks that the horse-cloth 'may be no more than the hunter's skin coat thrown over the back of the animal when led home by means of a halter made of thongs or withes to be there slaughtered.' If palæolithic man had advanced as far as Dr. Munro supposes, it was a short step to the domestication of the horse. It is hardly conclusive to say that, if he had tamed the horse, 'we would undoubtedly ere now have had an equestrian representation of the fact,' though it is also said that 'we have only as yet a preliminary instalment of these most interesting art productions.'[13] The representation may later be discovered. That palæolithic man, so far advanced as he was, was 'ignorant of the principles of religion,'[14] seems a hasty conclusion. If he had the beliefs of our Australians in such potent beings as Baiame, Nooreli, Daramulun, Mungun-ngaur, Pirmaheal, and Pundjel, that belief would leave no material traces, except, perhaps, the Bull-roarer, whose noise represents the voice of one or other of these beings. Now a small but unmistakeable pair of palæolithic bull-roarers in bone, or of amulets which are bull-roarers in miniature, one of them decorated with the sacred Australian pattern of herring-bone and concentric circles, have been found in a quaternary station in France.[15]

Palæolithic man in France, countless ages ago, was thus, especially if he had domesticated animals, immensely more remote from 'the beginning' than contemporary wild Australian tribes. They, again, with their copious languages, ingenious implements, complex institutions, and prolonged tribal assemblies, are infinitely in advance of those really primitive men among whom we must tentatively seek the origins of customary law regulating the family

---

[10] Munro, Archæological Journal, vol. lix. no. 234, p. 22.
[11] Ibid. p. 32.
[12] Ibid. p. 18.
[13] Ibid. p. 20.
[14] Ibid. p. 22.
[15] L'Anthropologie, Mars-Avril, 1902. For a brief bibliography of the bull-roarer see Mr. Frazer, The Golden Bough, iii. pp. 423-4, note 1.

and marital arrangements. A society almost incalculably ancient may have been much more advanced than a society of to-day, and the society of the lowest known modern savages must be equally advanced from the status of 'primitive man'.

The best proof of all that no Australians are now in or near 'the chrysalis state' of humanity, is to be found in their combinations into large friendly tribes, each covering a wide extent of country, and holding stated meetings, for social, political, religious, and commercial purposes. Mr. Matthews remarks on 'articles of barter,' exchanged 'at the great meetings which were held for the initiation of the youths of the tribes.' Among these articles were stone hatchets, first chipped, then ground, the tribes having passed out of the stage in which mere rude flaking sufficed. 'At the conclusion of the ceremonies, before the people dispersed, a kind of fair was held, when natives in whose country stone was plentiful, would barter their things with other people for reeds for making spears, rich plumage of birds, &c. ... or for any other articles brought by the various tribes for the purpose of exchange.'[16] We can scarcely conceive that this amount of tribal or inter-tribal unity was possible to man really primitive. Backward and conservative as the Australians are, we must not expect to find among them, with their highly complex customary laws, anything like the first beginnings of social regulations. To look for these, even among the naked and houseless hunters of Australia, is to organise failure in this research as to origins.

RECENT HISTORY OF THE SPECULATION AS TO THE EARLY HUMAN FAMILY

From the age of Aristotle onwards, inquirers naturally began with a belief in the Patriarchal Family as the original social unit. To this opinion, in a peculiar form, Mr. Atkinson returns, as will be seen. The idea was natural. Aristotle, like Hesiod, starts from 'the Man, the Woman, and the labouring ox,' though men and women were wedded long before oxen and other animals were domesticated. The Biblical account in Genesis opens with the same theory of the primal pair, whose children, brother and sister, must have married each other, as in the late Mr. Morgan's hypothesis of the 'Consanguine Family;' but, contrary to almost universal savage custom, and to Mr. Atkinson's 'Primal Law.'

---

[16] Journal and Proceedings Royal Society N.S.W., vol. xxviii. p. 305. See also Roth, Ethnological Studies, pp. 132-138. 1897.

In 1861, Sir Henry Maine's celebrated book, 'Ancient Law,' appeared. Herein he wrote that it was difficult to say 'what society of men had not been originally based on the Patriarchal Family.[17] His studies had lain chiefly in the law of civilised peoples, Romans, Hebrews, Greeks, Irish, and Hindoos; not in the customary law of the lowest races. He, like Mr. Freeman, concluded that the patriarchal family, by aggregation of descendants (and aided by adoption of outsiders, and by the ownership of the family by its Head), formed the gens, while the aggregation of gentes formed the tribe, and the aggregation of tribes made the State. But, as the gentes had traditions contrary to this theory, traditions of separate origins, he supposed that 'the incoming populace should feign themselves to be deduced from the same stock as the people on whom they were engrafted.' Thus we know that McUlrigs (Kennedys) of Galloway joined the remote Macdonnells of Moidart and Glengarry, and wore the Macdonnell tartan[18] (1745-1760), and so might come to pass as Macdonnells, though they still regard the Marquis of Ailsa, a Kennedy, as their chief, at least in Eilean Shona (Loch Moidart). In the same way the Camerons of Glen Nevis, though called 'Camerons,' were really MacSorlies, a branch of the Macdonnells, and from the sixteenth century to 1754 were always on ill terms with the chief of the clan Cameron, Lochiel. These are very modern instances, but illustrate Sir Henry's theory of incomers.

The members of the Roman tribes had traditions that they were not, really, of the same original blood with each other. Only by a fiction were they of the same blood. They did not all descend by natural increase from one patriarchal ancestor. There really did exist 'a variety of alien groups in a local tribe,' however they might all adopt the same name, and assert descent, in West Scotland from Somerled, let us say. This fact, of heterogeneousness within the 'tribe' among others, was so obvious and so imperfectly explained, by friends of the Patriarchal theory, that it occupied 'writers belonging to the school of so-called prehistoric inquiry,' as Sir Henry styled it.[19] They were not satisfied with the theory that Society arose in the Patriarchal Family, based on direct descent from, and ownership by, a single male ancestor. To be sure a Cameron will 'cross the hill,' and call himself Stewart, and a Chinese immigrant into Australia has discreetly entitled himself Alexander

---

[17] Ancient Law p. 132.
[18] Major Kennedy's portrait of 1750-1760 represents him in Macdonnell tartan. He was an agent of Prince Charles.
[19] Early History of Institutions, pp. 310, 311.

Mac-gillivray. But such accretions, and such legal fictions, do not explain the heterogeneousness of the local tribe, which, by the theory of some historians, is of common descent. 'Prehistoric inquirers' could not but notice that, among ruder 'non-Aryan' races of various degrees of culture, 'the family is radically different from the Patriarchal Family,' and suggests a different origin.

Roughly speaking, the groups of real or fancied kindred among various low races exhibit the peculiarity that the kin-name is often inherited from the mother, not from the father; that the maternal blood is stronger in determining such cases of inheritance as arise; and that marriage is forbidden within the recognised limits of the maternal kinship. It was natural for inquirers to derive this condition of affairs, this reckoning in the female line, from a state of society in which fatherhood (owing to promiscuity, or to polyandry—several husbands to one wife) was notably uncertain. Bachofen, who first examined the problem, attributed the system to a supposed period of the Supremacy of Women: McLennan to dubious fatherhood, and possible early promiscuity. The recovery of supremacy by men, or the gradual advance in civilisation, especially in accumulation of property, would finally cause descent to be reckoned through the male line, as among ourselves.

As to the question of early promiscuity—sexual relations absolutely unregulated—Dr. Westermarck, Mr. Crawley, and others have argued, and Mr. Atkinson argues, that it never existed, at least to any wide extent, and with any potent influence. We hear rumours of savages utterly promiscuous, say the Mincopies of the Andaman Islands, just as we hear of savages utterly without religion. But later and better evidence proves that the Andamanese have both wives and a God.[20]

Again, the lowest savages known are so far not 'promiscuous,' that they recognise certain sets of women as persons with whom (as a general rule, subject to occasional exceptions) certain sets of men must have no marital relations. It was the opinion of Mr. Darwin, as of Mr. Atkinson, that sexual jealousy, from the first, must probably have been a bar to absolute promiscuity, even among the hypothetical anthropoid ancestors of human race. To tell the truth, our evidence on these points, as to existing savages, is, as usual, contradictory.[21]

---

[20] Westermarck, History of Human Marriage, pp. 53-57.
[21] Mr. John Mathew declares that 'jealousy is a powerful passion with most aboriginal husbands' in Australia. Messrs. Spencer and Gillen, on the other hand, represent the aboriginal husband as one of the most complacent of his species, jealousy being regarded as 'churlish.' Messrs. Spencer and

WHAT IS EXOGAMY? DIFFICULTIES OF
TERMINOLOGY

In these inquiries a great source of confusion arises (as all students must be aware) from the absence of exact terminology, of technical terms with a definite and recognised meaning. Thus when my friend, the late Mr. John Fergus McLennan, introduced the word 'Exogamy,' in 'Primitive Marriage' (1865), he probably knew perfectly well what he meant. But he did not then, from lack of practice in an inquiry practically novel, and originated by himself, express his meaning with exactness. He at first spoke of exogamy as the rule 'which prohibited marriage within the tribe.'[22] But the word 'tribe' was later taken by Mr. McLennan to mean, and is now used as meaning, what cannot be a primitive community, a local aggregate of groups amicably occupying a considerable area of country; say the Urabunna tribe of Central Australia. Mr. McLennan did not wish to say that exogamy forbids an Urabunna tribesman to marry an Urabunna tribeswoman; he meant that exogamy prohibited marriage within the recognised kindred—that is, in this case, between members of totem kindreds of the same name, say Emu or Kangaroo. This fact he later made perfectly clear. But meanwhile such terms as 'horde,' 'tribe,' 'sub-tribe,' 'family,' 'gens,' 'section,' 'phratria,' 'clan,' many of them derived from civilised classical or Celtic usage, have been tossed up and down, in company with 'class,' 'division,' 'section,' and so on, in a way most confusing.[23] Odd new terms come from America, such as 'socialry,' 'tutelaries,' 'ocular consanguinity,' 'ethnogamy,' 'conjugal conation,' and so forth.[24] Most perplexing it is to find words like clan, family, tribe, gens, phratry, words peculiar to civilised peoples, Greek, Roman, or Celtic, applied to the society of savages. 'The term "clan" implies descent in the female line,' says the late Mr. Dorsey, following Major

---

Gillen are decidedly the better authorities. Mathew, Jour. Roy. Soc. N.S.W., xxiii. 404. Westermarck, p. 57. Native Tribes of Central Australia, p. 99.

[22] Studies in Ancient History, 1876, p. 41.

[23] The late Major Powell, of the American Bureau of Ethnology, used gens of a totem kin with descent in the male line, clan of such a kin with descent in the female line, and his school follows him. Mr. Howitt, on the other hand, uses 'horde' for a local community with female, 'clan' for a local community with male descent.

[24] 'The Seri Indians,' by W. J. McGee. Report of Bureau of American Ethnology, Washington, 1898.

Powell; but why take the Celtic term 'clan,' which has no such signification, and confer it on what is really a totem kindred with descent in the female line?[25] Next, 'several of the Siouan tribes are divided into two, and one into three sub-tribes. Other tribes are composed of phratries, and each sub-tribe or, phratry comprises a number of gentes.' Is there a distinction between the 'sub-tribes' of some tribes, and the 'phratries' of others, or not? Apparently there is not, but the method of nomenclature is most confusing.

I shall understand the terms which I employ, as follows:

The tribe, speaking of the Australians, for instance, is a large aggregate of friendly or not hostile human groups, occupying a territory of perhaps a hundred square miles, and holding councils and meetings for social and religious purposes. It is so far 'endogamous' that its members may marry within it—that is to say, it is no more endogamous than the parish of Marylebone. An Urabunna man, a man of the Urabunna tribe, may marry an Urabunna woman—if no special native law interferes. He may also at pleasure marry, out of his tribe, say a woman of the neighbouring Arunta tribe, again, if no special law bars the arrangement. So far the tribe, the large local aggregate of groups, stands indifferent. But, within the tribe, there are laws barring marital intercourse. First, each tribe is usually composed of two 'primary exogamous divisions,' or 'phratries,' so called; in the case of some tribes the phratries are named; for example, Matthurie and Kirarawa. Every man and woman, in such tribes, is either a Matthurie or a Kirarawa, and can only marry into the opposite division, and the children follow the name of the mother. These two divisions are called 'primary classes' by some students; 'phratrias' (from the Greek: Φρατρία) by others; 'sub-tribes' by others; or, again, 'moieties,' or 'groups.' I shall, in each instance, use the term ('class,' 'phratria,' 'moiety,' 'primary exogamous division,' 'group,' and the like) employed by the author whose opinion I am discussing, though I prefer 'phratry,' as 'class' has another significance; so has 'group,' &c.

Again, the tribe contains a number of totem kindreds (often called 'clans' or gentes, rather at random), that is, of sets of kin deriving their names from totems, plants, animals, or other objects in nature. To the possible origin of Totemism we return in a separate section. No Urabunna man may marry a woman of his own 'phratry,' nor of his own totem, and the children inherit the phratry

---

[25] 'Siouan Sociology,' Report of American Ethnological Bureau, 1897, p. 213.

and totem names from the mother. Finally, there are sets of relationships, roughly indicating, it would seem, seniority by generations, and degrees of actual or supposed kindred. Within many of these, which I shall style 'classes' (they have other terms applied to them), marriage is forbidden. Thus there are bars of three several sorts on the intermarrying of an Urabunna man with an Urabunna woman. In a way, there are three grades of exogamous prohibitions.

Mr. McLennan, who introduced the word 'exogamy,' defined it thus: 'an exogamous marriage is a marriage between persons of different clans of kinship, not entered into fortuitously, but because of law declaring it to be incest for a man to marry a woman of his own clan.'[26] The same community cannot be 'both exogamous and endogamous,' as some suppose. Thus Lord Avebury writes, 'some races which are endogamous as regards the tribe, are yet exogamous as regards the gens.' But really 'exogamy is the law prohibiting marriage between persons of the same blood or stock as incest— often under pain of death—and endogamy is the law prohibiting marriage except between persons of the same blood or stock.'[27] In Mr. McLennan's sense I shall take the word 'exogamy,' while dealing with peoples apparently nearest the beginning.

Later, when descent in the male line is established, the prohibition on marriage within the totem name comes to apply, sometimes, to marriage within the local district held by the men of the name. The old prohibition, we see, is to many within the recognised limit of the blood kinship, or stock, designated by the totem name. But, as tribes advance to kinship through males, and as, thereby, groups of one totem name come to possess one region of country, it often happens that exogamy prohibits marriage between persons dwelling in that region. Whereas Grouse was forbidden to marry Grouse; later, the Grouse living together, say in Corradale, the exogamous prohibition takes the shape 'persons dwelling in Corradale must marry out of Corradale.' The name marking the exogamous limit is now, in such cases, local, but the prohibition is derived from the older tabu on marriage between 'persons of the same blood or stock'—all those in Corradale being conceived to share the same blood or stock. This origin of 'local exogamy' must be kept in mind, otherwise confusion will arise. There are a few cases, even in Australia, where even local exogamy

---

[26] Studies in Ancient History, second series, p. 265.
[27] Studies in Ancient History, second series, p. 46. In an appendix to Mr. Morgan's Ancient Society, Mr. McLennan's terms are severely criticised.

has become obsolete, and marriage, as with ourselves, is prohibited between persons of near kindred simply.

Now, if I may venture to interpret the mind of Mr. John Fergus McLennan, I conceive that he regarded the totemic division as older than the 'phratry' or the 'class' bar, and he thought it the oldest traceable exogamous limit. Not to marry within the totem name (no male Emu to marry a female Emu) was, in Mr. McLennan's opinion, the most archaic marriage law.[28] This appears from the words of Mr. McLennan's brother, Mr. Donald McLennan.[29] He writes: 'As the theory of the Origin of Exogamy took shape, and the facts connected reduced themselves to form in his mind, the conclusion was reached that the system conveniently called "Totemism" ... must have existed in rude societies, prior to the origin of Exogamy.[30] This carried back the origin of Totemism to a state of mind in which no idea of incest existed. From that condition my brother hoped to trace the progress of Totemism— necessarily a progress upwards—in connection with kinship and Exogamy. It may here be said that he had for a time a hypothesis of the origin of Totemism, but that he afterwards came to see that there were conclusive reasons against it.'

Meanwhile may we not, then, assume that, in Mr. McLennan's opinion, the earliest traceable human aggregate within which matrimony was legally forbidden was the totem kin, indicated by the totem name, the totem tabu, and the totem badge, or symbol— where it existed?

We now see how heterogeneous elements came to exist in the tribe of locality, a puzzle to the friends of the theory of the Patriarchal Family. For the nature of totemism, plus exogamy and female descent, is obviously such that under totemism, each family group even (each 'fire circle' of men, wives, and children), must contain persons of different totems. The father and mother must be of different totems (persons of the same totem not intermarrying), and the children must inherit the totem either of the father or of the mother.[31] When paternal kinship is not only recognised (as, in practical life, it always is), but becomes exclusive in its influence on

---

[28] I shall call each set indicated by a totem name a 'totem group,' if the members live together; a 'totem kin,' if they are scattered through the tribe.
[29] The Patriarchal Theory, pp. 6, 7, 1885.
[30] Meaning by Exogamy, not a mere tendency to marry out of the group, but a customary law with a religious sanction.
[31] Here the unusual case of the Arunta offers an exception to the rule; a point to be discussed later.

customary law, and when an approach to the Patriarchal Family, with the power of the patriarch, is evolved, all the members of the family in all its branches will (if Totemism persists) have the same totem; derived from the father. Thus there will now be a local totem group, a group mainly of the same totem name, as is practically the case in parts of Central Australia.[32]

It is necessary to understand this clearly. Take a very early group, in a given district; suppose it, at first, to be anonymous, and let it later be called the Emu group. So far, all members of the group will be Emus, they will form an Emu local group. But, next, suppose that there are many neighbouring groups, also at first anonymous; let them later be styled Rat, Cat, Bat, Sprat. Suppose that each such group now (for reasons to be indicated later) takes its wives not from within itself, but from all the other groups; that these women bring into the Emu group their group names; and that their children inherit their names from their mothers. Then the name, 'Emu group,' will cling to that local aggregate, as such; but, in time, the members of the Emu group will all be, say, Rats, Cats, Bats, and Sprats, so called from the group-names of their alien mothers. Suppose that, for one reason or another, children at last come to inherit their names and totems from their fathers. Then a Cat father will have Cat children, though his wives may still be of different totems, and his sons' children will also be Cats, and so the local group will become mainly, if not wholly, a group of one totem, the Cat. The Arunta of Central Australia do trace kinship in the male line, and thus there is 'one area which belongs to the Kangaroo men, another to Emu men, another to Hakea flower men,' and so on. This has reached such a pitch that 'in speaking of themselves the natives will refer to these local groups,' not by the prevalent totem names in each, but 'by the name of the locality which each of them inhabits,' namely, as men of the Iturkawura camp, and so on.[33] Thus we might say 'the Glen Nevis men,' 'the Corradale men,' and so on.

Thus we begin with an anonymous group, or group of unknown name, a local group. We introduce Totemism, and that group becomes a local group with a totem name. Granting exogamy (prohibition of marriage within the group), and reckoning in the female line, it soon develops into a local group made up of various totems, but, at first, as a local group, it probably retains its original totem name among its neighbours. Reckoning, still later, through the male line, we again meet, as at first, a local totem group, but already Totemism is on the wane, and the groups are soon to be

<hr>

[32] Spencer and Gillen, pp. 8-10.
[33] Ibid. pp. 8-9.

12

called by the territorial names of their lands. At this stage totem names are tending to decay, and the next step will probably be to style the group by the name of some remembered, or mythical, male ancestor, such as 'children of Donald'—Macdonalds.

Thus if, at a given time, the name of a certain male ancestor is substituted, as 'eponymous,' for the totem name, or the district name, we shall find a local group of, say, Sons of Donald, into which other groups, Sons of Sorlie, or Ulrig, will enter, as occasion serves, and be more or less absorbed. A State may at last arise, say, 'Softs of Israel.'

We are not assuming, however, that all human societies have passed through the totemistic and exogamous stages.

TOTEMISM AND EXOGAMY

But what was the original unit, the totem group, or other division outside of which alone could marriages be arranged? And why was the totem name the limit? Returning to Mr. Donald McLennan's account of the opinions which his brother did not live to set forth, Totemism arose 'in a state of man in which no idea of incest existed.' On this theory, I presume, there would be totem groups before exogamy arose; before it was reckoned 'incest' to marry within the totem name. This, as we shall see, appears to be sometimes the opinion of the best Australian authorities, Messrs. Fison and Howitt, and Messrs. Spencer and Gillen. It is also the theory of Arunta tradition. The totem belief, as it now exists, imposes many tabus: you may not (as a rule) kill, eat, or use the plant or animal which is your totem; still less perhaps, in the long run, may you 'use,' sexually, a woman of your totem. If this, or a kindred totem tabu, is the origin of exogamy, then to exogamy (as a law, though not necessarily as a tendency) the totem is prior in time. But I have no reason to suppose that Mr. McLennan ever regarded the totem tabu as the origin of exogamy. In his published works he offers another theory, not commonly accepted.

But the important thing to note is that exogamy may conceivably (contrary to Mr. McLennan's opinion, but in accordance with that of Mr. Atkinson) have existed, or rather tended to exist, before totems arose; much more, then, previous to the evolution of totem names, of totem tabu, and of the idea of incest, as a sin, or mystic misdeed, and as an offence to the totem—a religious offence to God, or to ancestral spirits. Persons may have been forbidden to

13

marry within their local group, their 'fire circle' before that group had a totem, or a totem name, and they may have been forbidden for reasons purely secular, to which the totem later lent a sanction, and a definite limit. Thus Mr. Tylor, our most sagacious guide in all such problems, writes 'Exogamy can and does exist without Totemism, and for all we know was originally independent of it.'[34]

It is part of my argument that exogamous tendencies, at least—that is, a habit of seeking female mates outside of the fire-circle—may very well have prevailed before any human group had even a totemic name. But exogamous tendencies are not, of course, the same thing as exogamy strictly defined, and sanctioned by religious or superstitious fear, and by secular penalties inflicted by the tribe. Against the notion that exogamy may have been prior to Totemism, Mr. Robertson Smith argued that very early man would not be restrained from marriages by such an abstract idea as that of kindred—'not to marry your near kin'—while the idea of kindred was still fluid, and not yet crystallised around the totem name.[35] But, without thinking of kindred by blood, perhaps without recognising consanguinity (though it must have been recognised very soon), early man may have decided that 'thou shalt not marry within this local group or crowd, of which I am head.' Nothing abstract in that! There was no tribal law—there were as yet (I suppose) no tribes—only the will of the head of each small set of people practically enforced exogamy.

We can have no certainty on this point, for we know of no pre-totemic race, no people who certainly have not yet entered into the totemic stage. Any such people, probably, in the remote past, had no idea of incest as a sin, or of exogamy as a law sanctioned by a tabu. But they may have, at least, had a strong tendency to marry outside of the circle of the hearth, the wandering hearth of homeless nomads ranging after food.

The reader of Mr. Atkinson's treatise will find that this kind of exogamy—marriage outside the local group—would, on his theory, be the rule, even when no idea of blood kindred, or of incest as a sin, need have arisen; and no totem, or anything else, had yet been named. The cause of the prohibition would, in Mr. Atkinson's opinion, be the sexual jealousy of the hypothetical patriarchal anthropoid male animal; and, later, the sexual jealousy of his adult male offspring, and of the females. Still later the group, already in practice exogamous, would accept the totem name, marking off the group from others, and the totem name, snipe, wolf, or what not,

[34] 'Remarks on Totemism,' Jour. Anthrop. Inst., August, November, 1898.
[35] Kinship in Early Arabia, p. 187.

14

would become, for the time, the exogamous limit. No man and woman of the same totem name could intermarry. Still later, a myth of kinship with the totem would arise, and would add the religious sanction of a tabu.

A prohibition may perhaps have arisen very early, even if Mr. Atkinson's hypothesis (that the rule of marriage outside the group arose in a state of brutality) be rejected. 'The origin of bars to marry is, in fact, complex,' writes Mr. Crawley. A dislike of marriage with a group-mate, familiar, through contiguity, from infancy, may have been developed among early men;[36] and may have been reinforced by the probably later superstitions which create 'sexual tabu,' and mutual avoidance, among many existing peoples. Men and women are, by savages, conceived to be mysteriously perilous to each other, especially when they live in close contiguity. Mr. Crawley also allows for Mr. Atkinson's main factor, jealousy, 'proprietary feeling, which is one crude means by which the family has been regulated and maintained.'[37] If these things were so (whether we go back to Mr. Atkinson's semi-brutal ancestors, or not), then, contrary to Mr. Donald McLennan's opinion, and to general opinion, it would not 'appear to be possible to demonstrate that Totemism preceded exogamy,' or at least preceded the exogamous tendency. For, in the first place, exogamy might conceivably tend to arise before the explicit idea of kinship—whether male or female—arose. Mr. Atkinson's 'primal law' would be unuttered in speech (speech, by his theory, there was none), but would amount to this: 'I, the patriarchal bull of this herd, will do my best to kill you, the adult young bulls, if you make any approaches to any of the cows in this crowd.' There is no notion of 'incest,' but there is jealousy producing the germ exogamy. The young bulls must find mates outside of the local herd—or do without. This rule persisted, on Mr. Atkinson's theory, till the hypothetical anthropoid became a man, and named his group (or had it named for him, as I later suggest) by a totem name.

But real human and speaking beings might enforce marriage outside of the group, though they did not perhaps think explicitly of kindred (or, at least, did not think the idea fully out), still less of 'incest,' as sin. Mr. McLennan's theory, as given in his works, was partly identical with that of Mr. Atkinson. 'The earliest human groups can have had no idea of kinship'—they must, therefore, have been rather low savages. 'But,' he said, 'they were held together by a

---

[36] But, as Dr. Durkheim says, man and wife might soon abandon each other, if familiarity breeds contempt.
[37] Journal of the Anthropological Institute, May, 1895, p. 444.

feeling of kinship,' not yet risen into explicit consciousness. Cat and kitten have, probably, feeling of kinship, and that feeling is very strong, while it lasts, in the maternal cat, while between semi-human mothers and children, arriving so very slowly at maturity, mother-kin must have been consciously realised very early. Mr. McLennan then showed the stages by which the savage would gradually, by reflection, reach explicit consciousness of female kinship, of mother-relationship, sister and brother relationship, and all the degrees of female kin.

But Mr. Fison and others have argued powerfully against this theory.[38] Moreover, we find male relationships, as we saw—'descent counted in the male line'—among the Arunta of Central Australia, whom Mr. J. G. Frazer regarded, in 1899, as actually 'primitive;' while the neighbours of the Arunta, the Urabunna, reckon through the female line.[39] Mr. Crawley, for various reasons, says, 'the famous Matriarchal theory' (the prepotency and dominion of women) 'was as exaggerated in its early forms as was the Patriarchal.... It is a method of tracing genealogy, more convenient in polygamous societies and more natural in primitive times when the close connection of mother and child during the early days of infancy emphasises the relation.'[40] Dr. Westermarck argues to a similar effect.[41] His motive is to discredit the theory of promiscuity, and consequent uncertainty of fatherhood, as the cause of reckoning on the spindle side. But the Arunta, who reckon on the sword side, actually do not even know that children are the result of sexual intercourse, according to Messrs. Spencer and Gillen. How they can have any idea of blood-kinship at all is, therefore, the mystery. It may perhaps be argued that they have none. But these ignorant Arunta reckon descent through the male line—while the Royal Picts, in early Scotland, infinitely more civilised, reckoned by the female line.

For myself, I still incline to the opinion[42] that the reckoning of descent through the woman is the more archaic method, and the method that, certainly, tends to dwindle and disappear, as at last it did among the Picts. This applies to human society, not to that of Mr. Atkinson's hypothesis, in which the question is not of kin, but of property. 'Every female in my crowd is my sole property,' says—or feels—Mr. Atkinson's patriarchal anthropoid, and the patriarch

---

[38] Kamilaroi and Kurnai, p. 132. 1880.
[39] Spencer and Gillen, p. 70. Frazer, Fortnightly Review, April, May, 1899.
[40] The Mystic Rose, p. 460.
[41] History of Human Marriage, pp. 105-113.
[42] Tylor, J. A. I. xviii. 3, 254.

gives expression to his sentiment with teeth and claws, if he has not yet learned to double up his fist, with a stone in it. 'These were early days.'

THEORIES OF EXOGAMY. MR. MCLENNAN'S THEORY

In any case, Mr. McLennan's hypothetical first groups, like Mr. Atkinson's, were very low indeed. They developed exogamy, not (as in Mr. Atkinson's theory) through sexual jealousy on the part of the sires, but, first, through regular female infanticide. This practice, being reasonable, could not prevail among Mr. Atkinson's anthropoids.[43] Girl babies being mostly killed out, women became scarce. Neighbouring groups being hostile, brides could only be procured by hostile capture. Each group thus stole all its brides and became exogamous, and marriage inside the group became a sin, by dint of 'a prejudice strong as a principle of religion.'

This theory of Mr. McLennan's is, I think, quite untenable. The prevalence of female infanticide, at the supposed very early stage of society, is not demonstrated, and did not seem probable to Mr. Darwin. Even if it existed, it could not create a prejudice against marrying the few women left within the group. Mr. McLennan, unhappily, was prevented by bad health, and death, from working out his hypothesis completely. His most recent statement involves the theory that the method of the Nairs of Malabar, living in polyandrous households (many men to each woman) was the earliest form of 'marriage.' But people who, like the Nairs, dwell in large households, are far indeed from being 'primitive.' 'A want of balance between the sexes' led, Mr. McLennan held, to 'a practice of capturing women for wives,' and was followed by 'the rise of the law of exogamy.' The first prohibition would be against capturing women of the kindred (marked by the totem), for such capture, if resisted, might involve the shedding of kindred blood. Women being scarce, through female infanticide, kindred groups would not give up or sell their women to each other (though to the males of the groups, such women could not be wives), nor could women be raided from kindred groups, as we saw. So they would be stolen from alien groups, 'and so marriages with kindred women would tend to go into desuetude.' The introduction of captured alien wives would change the nature of matrimonial relations. Under the Nair system 'a woman would live in the house of her mother, and under

---

[43] The practice however, is attributed to tame canary birds.

17

the special guardianship and protection of her brothers and her mother's brothers. She would be in a position of almost absolute independence of her husbands....'

But really pristine man and woman can have had no houses, no matriarchal rule of women. The Nairs, not being primitive, have houses, and their women have authority: pristine man was not in their condition. However, captured alien wives would, Mr. McLennan argues, be property, be slaves; and men would find this arrangement (now obsolete) so charming that polyandry and the reign of woman would go out. The only real legal marriage would be wedlock with an alien, a captive, a slave woman. Marriage with a woman of the same stock would be a crime and a sin. It would be incest.[44] Really it would be, at worst, concubinage.

This theory seems untenable at every point, community of wives, female infanticide, household life, supremacy of women in the household, living with a non-captive wife reckoning as incest, and, in short, all along the line. Even if the prejudice against marrying native women did exist, it could not be developed into the idea of sin—granting that the idea of sin already existed. To be sinful, endogamy within the group must have offended some superstitious belief, perhaps the belief in the totem, with its tabu.[45]

MR. CRAWLEY'S THEORY

To disengage from his learned book, The Mystic Rose (1902), Mr. Crawley's theory of the origin of exogamy is no easy task. He strongly insists on the 'religious' element in all early human thought, and as in 'religion' he includes the vague fears, misgivings, and ideas of 'luck,' which haunt even the least religious of modern men, we may say that 'religion,' in this sense, mingles with the thought of all ages. The present writer, like Dr. Johnson, is an example of the 'religious' character, and of Mr. Crawley's remark that 'human nature remains potentially primitive.' To the 'religious' man or woman (using 'religious' in this sense) the universe is indeed a thing of delicate poise, and may 'break, and bring down death,' if we walk under a ladder, or spill the salt, or enter a doorway with the wrong foot foremost, or fail to salute a magpie, or the new moon. The superstitious anthropologist, of course, knows that all these apprehensions of his are utterly absurd, but the savage is careful and troubled about them. The Philistine, on the other hand, is

---

[44] Studies in Ancient History, second series, pp. 57-65.
[45] Cf. Custom and Myth (A. L.), p. 258.

proud of his conquest of these airy terrors: he 'cannot imagine what people mean by such nonsense,' and, exactly so far as he is sincere, he cannot comprehend early mankind.

Now, as to exogamy, our difficulty is to understand why breach of the rule against certain marriages is, everywhere, so deadly a sin: so black an offence against 'religion.' Mr. Crawley's explanation is not, perhaps, easily to be disengaged from the mass of his work, but it begins in his appreciation of the δεισιδαιμονία of early men, their ever-present sense of 'religious' terrors. 'Thus all persons are potentially dangerous to others, as well as potentially in danger....'[46] This sense of peril arises 'in virtue simply of the distinction between a man and his fellows.' Much more, then, are women dangerous to men, and men to women, the sexes being so distinct from each other. We know that the most extraordinary precautions are taken to avoid contact with women in certain circumstances, and a well-known story of Sir John Mandeville's is only one case of the fact that the bridegrooms of some races, from a superstitious terror, insist on being made cocus en herbe. Messrs. Spencer and Gillen give the instance of 'the marriage ceremony' (an odious brutality) among the Arunta of Central Australia.[47] It is perhaps intended to deliver the bridegroom from a peril imagined by superstition (as in Mandeville's tale);[48] and, without it, the Australian would resemble the man derided in the old Scottish song:

> The Bridegroom grat when the sun gaed doon.

Thus a 'religious' dread attaches among savages (the theory holds) to all marriages; all are novelties, new steps in life, and therefore are so far 'sinful' that they involve a peril, vague but awful, the creation of superstition. Marriages contrary to the exogamous rule, are only especially and inexplicably bad cases of the 'sin'—that is, mystic danger—of marital relations in general, as I understand Mr. Crawley. Marriage ceremonies of every kind are devised to avoid 'sin,' as our Marriage Service candidly states, using 'sin' in the Christian sense of the word. But there are savage marriages, those forbidden by the law of exogamy, which, as a general rule, no ceremony can render other than sinful. So great and terrible is the danger of such marriages—namely, among many savages, between persons of the same totem, that it threatens the whole community,

---

[46] Mystic Rose, p. 31.
[47] Spencer and Gillen, pp. 92-93.
[48] Lord Avebury's view that the 'rite' implies compensation to the other males of the community will be considered later.

just as the marriage of Charles I. with a Catholic bride caused the Plague, according to the Rev. Mr. Row, and therefore such unions are punished by the death penalty, and are but seldom left to the automatic vengeance of the tabu. Foremost in this black list of sins are the unions of brothers and sisters of the full blood, though, we must remember, these are not more heavily punished than marriage between a man and woman of the same totem, even if the pair come together from opposite ends of the continent, and are not blood relations at all. Why is this?

As I understand Mr. Crawley, the sexes, in savagery, avoid each other's society in everyday life, partly from 'sexual tabu'—the result of the superstitions already indicated; partly because of 'sexual solidarity,' perhaps even of 'sexual antipathy.' In fact, men and women are often very much in each other's way. We do not want women in our clubs and smoking-rooms—nor do savages—and we despise a man who lurks in drawing-rooms when his fellows are out of doors; a man who is a pillar of luncheon parties and of afternoon tea. But this separation of the sexes is especially rigid between the children of the same hearth, even among nomads. The boys go with the father, the girls with the mother. The manlike apes have the same ideas. 'Diard was told by the Malays, and he found it afterwards to be true, that the young Siamangs, when in their helpless state, are carried about by their parents, the males by the father, the females by the mother.' 'The nests ... are only occupied by the female and young, the male passing the night in a fork of the same tree or another tree in the vicinity.'[49]

These facts of ape etiquette would, to use an Elizabethan phrase, have been 'nuts' to Mr. Atkinson, and prove that sexual separation of the children is a very early institution. In Australia, New Caledonia, and other countries, brothers and sisters must not even speak to each other, and must avoid each other utterly. Thus the danger and 'sin' of the most innocent intercourse between brothers and sisters is emphasised; much more awful, then, are matrimonial unions of brother and sister. 'The extension' (of this idea) 'by the use of relationships produces the various forms of exogamy,' says Mr. Crawley.[50] There are difficulties here; for example, Mr. Crawley tells us that incest did not 'need prevention,' though the rules of brother-and-sister avoidance seem really to mean that it did, or was thought to do so (but perhaps only superstitious dread of ordinary intercourse caused the rule?), and

---

[49] Westermarck, p. 13. Citing Brehm, 'Thierleben,' i. 97, Proceedings R.G.S. xvi. 177.
[50] Mystic Rose, p. 443.

though we know of regions where such incest, in early youth, is said to be universal.[51] 'Such incest,' says Mr. Crawley, 'is prevented by the psychological difficulty with which love comes into play between persons either closely associated, or strictly separated before the age of puberty....'[52] Now we know that lust does come into play—for example, among the Annamese—between brothers and sisters not closely separated; and we also know that, the more persons are 'strictly separated,' the more does the novelty and romance, when they do meet, produce natural attraction, as between Romeo and Juliet. Incest among the young is really prevented by the religious horror with which, by most peoples, it is regarded; as well as, among the civilised, by the constant and sacred familiarity of family life. The bare idea of it can only occur, as a desirable notion, to a boyish revolutionary, like Shelley, or to minds congenitally depraved.

Again, men and women of the same totem have no 'avoidances' forced upon them, as far as I know (and, as they may not marry, this is an oversight); yet their marriages are as terribly sinful as marriages between brother and sister of the full blood. Mr. Crawley writes, 'Obviously the one invariable antecedent in all exogamous systems, indeed in all marriage systems, is the prohibition of marriage "within the house."' But, we reply, A (a male) and B (a female), of the same totem, may never have been in the same house, or in the same degree of latitude and longitude, before they met and fell in love. As to 'house,' houses they may have none. Yet their union is a deadly sin. Mr. Robertson Smith is said to have 'set the question in the right direction,' when he wrote, 'whatever is the origin of bars to marriage, they certainly are early associated with the feeling that it is indecent for house-mates to intermarry.'[53]

But what is early need not be primary.

Again, if Mr. Crawley reads on, he will find, I think, that the context of Mr. Robertson Smith's argument shows him not to have held that exogamy arose in 'the feeling that it is indecent for "house-mates"' (or tent-mates) 'to marry.' For Mr. Robertson Smith adds, 'it will not do to turn this argument round, and say that the pre-Islamic law of bars to marriage may have arisen ... in virtue of a custom that every wife and her children shall have their own tent. In any case, we cannot speak of 'house-mates' before there were houses. But if for 'house-mates' we read 'hearth-mates,' then no

---

[51] Westermarck, p. 292.
[52] Mystic Rose, p. 222.
[53] Kinship and Marriage in Early Arabia, p. 170.

sense of 'indecency,' as on Mr. Crawley's theory, need necessarily attend their marriage, for hearth-mates may be of different totems, derived from different mothers, and may be marriageable enough, at least as far as totem law is concerned. A, male, an Emu, marries B, a Bandicoot, and C, a Grub. His children by B have the Bandicoot totem, his children by C have the Grub totem. As far as totem law goes, these children may intermarry, but this is not allowed in practice to-day. Mr. Mathews says, of the Kamilaroi, 'in order to prevent such a close marriage' (of brother and sister on the father's side), 'every tribe has strict social customs, founded upon public opinion, which will not tolerate the union of a man with a woman whose blood relationship is considered too near.'[54] Australian ethics, long trained under the old totem and phratry prohibitions, are now sufficiently enlightened to reject unions which we also forbid. But it cannot have been so in the beginning, or the totem and phratry tabus on marriage would have had no occasion to exist. It would have sufficed to say, 'Thou shalt not marry thy sister, or mother,' and the totemic rule would have been a cumbrous superfluity. Superfluous it would have been, even under the hypothetical 'group marriage system,' where the law would have run 'Thou shalt not marry thy group-sister or group-mother.'

While Mr. Matthews gives a kind of bye-law, forbidding marriage, under female descent, with the paternal half-sister, Mr. Fison avers that the Kamilaroi do allow such unions. 'It is marriage within a phratria,' but not within a totem.[55] The fact was denied, or at least questioned, by many correspondents, but Mr. Fison believed it to be authentic. 'The natives justified it on the ground that the parties were not of the same mudji' (totem). Apparently these natives, who let a man marry his father's daughter, had not arrived at an objection to unions of 'too near flesh.' But mere decadence, under European whisky, may be the explanation. Mr. Matthews denies, as we saw, what Mr. Fison asserts, as to the Kamilaroi. Mr. Crawley writes, 'if we apply to the word "indecent" the connotation of sexual tabu ... and if we understand by "house-mates" those upon whom sexual tabu concentrates, we have explained exogamy.'[56]

Scarcely, for sexual tabu against marriage, in fact, now, at least, concentrates on people of near kin, and on totem-mates, man and woman of the same totem, and they may be 'house-mates,' or 'hearth-mates,' or they may not (in polygamous society), and the hearth-mates (as far as the totem rule goes, but not now in practice)

---

54 Proc. Roy. Soc. N.S.W. xxxi. 166.
55 Kamilaroi and Kurnai, pp. 42,46, 47, 115.
56 Mystic Rose, p. 443.

may thus be intermarriageable, as not of the same totem, while totem-mates, from opposite ends of a continent, are not intermarriageable (except in the peculiar case of the Arunta and cognate tribes).

But Mr. Crawley may reply that each totem, originally, did really pertain to all members of each small local group, and that the totem prohibition was extended, later, to all groups of the same totem name, however distant in space. Thus according to the Euahlayi blacks there were originally no totem names, but the divine Baiame gave them to mortals with the rule that no pair of the same totem name were to marry, 'however far apart their hunting grounds.' Thus considered, the tabu which forbids an Emu man to marry an Emu woman, would mean no more, originally, than that marriage between persons living in the close contiguity of the same local group (in this case the Emu group) was forbidden. There might be no original intention of prohibiting marriage with a person of an Emu group, dwelling a thousand miles away; probably no such group was known to exist. The original meaning of exogamous law, I repeat, would be merely 'you must not marry a hearth-mate,'—or a 'house-mate,' in Mr. Crawley's phrase—the hearth-mates, in this particular instance, being delimited by the name 'Emu.' So far my conjecture agrees with that of Mr. Crawley. The extension of the prohibition to persons of the same totem-name, however remote their homes and alien their blood, I am content to regard as a later kind of accidental corollary. There came to be totem kins of the same name, far remote, and thus, as it were casually, the law acquired an unpremeditated sweep and scope, including persons not really of the same group or blood, only of the same name.

But why was there originally any objection at all to marrying the most accessible bride, the female hearth-mate? Here, as I have tried to show, Mr. Crawley would explain by his idea of sexual tabu. All men are regarded with superstitious dread by all women, and vice versa; above all, as a daily danger, the men, or women living in close contiguity must avoid each other. To keep them apart all sorts of tabus and avoidances are invented, including the tabu on their marriage.

This is a plausible and taking theory, and I am far from arguing that it cannot be a true theory. But the insuperable difficulty of deciding arises from the circumstance that we know nothing at all about the intellectual condition of the more or less human beings among whom the prohibition of marriage within the group first arose. Were they advanced enough to be capable of such a superstitious dread of each other as the supposed cause of the prohibition takes for granted? Males and females, among the lower animals, have no such superstition. It requires human imagination. On the other hand, animal jealousy was well within their reach, and Mr. Atkinson derives the original prohibition of marriage within the group from the sheer sexual jealousy of the animal-patriarch. In his opinion the consequent aversion to such wedlock crystallised into a habit, as the race advanced towards full humanity.

Even before his anthropoid clients were completely human, the group would be replete with children of females not of the full group blood, captives, and therefore these children (if blood kin through females were regarded) would be eligible as wives. But this would not yet, of course, be understood. Perhaps it would not be fully understood till the totem name was given to, and accepted by, each group, and so there was a definite mark set on each woman brought in from without the group, and on her children, who bore her totem name. After that, each totem group obviously contained members of other totems, and those, being now recognised by their mother's totem names, were technically intermarriageable. What had been a group not explicitly conscious of its own heterogeneous elements, became, in fact, an assemblage of recognised heterogeneousness, capable of finding legal brides within itself, and no longer under the necessity (had it understood) of capturing brides from without in hostile fashion. Such an assemblage would, or might, come to consist of families, dwelling, or rather wandering, within a given region, all on terms of friendship and mutual aid. I take it that, by this time, improved weapons and instruments, and improved skill, enabled groups larger than the small original groups to live in a given area. In fact, the group would, or might, be a small local 'tribe,' but, probably, was unconscious of the circumstance. If conscious, one cause of hostility among the groups was at an end, there was no necessity for stealing women, a system of peaceful betrothals within the group might now arise, though certain facts, to be dealt with later, raise a presumption, perhaps, that this relatively peaceful state of life did not appear until two of the original local totem groups coalesced in connubium, intermarrying with each other, in fact becoming 'phratries.'

To produce the new condition of affairs, two factors were necessary: first, a means of distinguishing the captured women within every group from each other, and from the group into which they were brought by capture. This means of distinction was afforded by the totem names. Next, a recognition of kinship was needed, and this was supplied, let us conjecture, by naming the children of each of the captive women after the totem name of the group from which she was captured. If all the children

24

indiscriminately were called by the totem name (say Emu) of the local group into which their mothers had been brought—that is, by the totem name of their fathers—there would be no recognisable heterogeneity within that group, and so there would be, within the local group, no possible wives, under the exogamous rule. Whether polyandry then existed, or not, still all the fathers were of one local totem name, say Emu, and children could only be differentiated by styling them after the totem names of their alien mothers. This is usually done among the savages who are least advanced, but not among the Arunta, whose totem names, as we shall see, by a curious divergence, do not indicate stock, but are derived from a singular superstition about ancestral spirits, of various totems, incarnating themselves in each new-born child.

Mr. McLennan, in Primitive Marriage (1865), had arrived at conclusions very like these. The primitive groups 'were assumed to be homogeneous.... While as yet there was no system of kinship, the presence of captive women in a horde' (group), 'in whatever numbers, could not introduce a system of betrothals'—the women and their children not yet being differentiated from each other, and from the group in which they lived. Mr. McLennan, in 1865, did not ask how these women ever came to be distinctly differentiated, each from each, and from the group which held them, though that differentiation was a necessary prelude to the recognition of kindred through these women. But presently, in his Studies of Totemism (1869), he found, whether he observed the fact or not, the means of differentiation. Differentiation became possible after, and not before, each primitive group received a totem name, retained by its captive women within each group to which they were carried, a name to be inherited by their children in each case.

He says, 'heterogeneity as a statical force can only have come into play when a system of kinship led the hordes to look on the children of their foreign women as belonging to the stocks of their mothers.' That was impossible, before the totem or some equivalent system of naming foreign groups arose, a circumstance not easily observed till Mr. McLennan himself opened the way to the study of Totemism.[57]

It thus appears that Mr. Crawley's theory of exogamy and mine are practically identical in essence (if I rightly interpret him). The original objection was to the intermarriage of the young of the group of contiguity, the hearth-mates. If there was but one male of the elder generation in the group of contiguity, these young people would be brothers and sisters. If there were two or more males of

---

[57] See Studies in Ancient History, pp. 183-186.

the elder generation, brothers, the group would include cousins, who (even before the totem name was accepted by the group) would also be forbidden to intermarry. When the totem name was accepted, cousins, children of brother and sister, and even brothers and sisters, children of one father, by, wives of different totems, would be, technically, intermarriageable: though their marriages may, in practice, have been forbidden because they were still of the group of contiguity, and as such bore its local totem name, say, Emu, while, by the mother's totem name, they may have been Bats, or Cats, or anything. Where I must differ from Mr. Crawley is in doubting whether at this hypothetical early stage, the superstitions which produce 'sexual tabu' had arisen. We cannot tell; but certainly, as soon as the totem name had given rise to the myth that the totem, in human beings as in animals and plants, was inviolable—the beast or plant of the totem blood not to be killed or eaten,[58] the woman of the totem name not to be touched—so soon would endogamy, marriage within the totem, be a sin, incest. This it would be; the totem tabu once established, whether sexual tabu, or sexual jealousy, or both, caused the first prohibition, not to marry group, mates. Here we may briefly advert to Dr. Westermarck's theory of exogamy, though it interrupts the harmonious issue of our speculations.

DR. WESTERMARCK'S THEORY

As to exogamy, Dr. Westermarck explains it by 'an instinct' against marriage of near kin. Our ancestors who married near kin would die out, he thinks, and they who avoided such unions would survive, 'and thus an instinct would be developed,'[59] by 'Natural Selection.' But why did any of our ancestors avoid such marriages at all? From 'an aversion to those with whom they lived.' And why had they this aversion? Because they had an instinct against such unions. Then why had they an instinct? We are engaged in a vicious circle. 'Lastly it is not scientific to use the term instinct of this kind of thing.'[60]

---

[58] This is the view of Dr. Durkheim, who explains the blood superstition. Cf. Reinach, L'Anthropologie, x. 652.
[59] History of Human Marriage, p. 352.
[60] Compare Mr. Crawley, Mystic Rose, pp. 444-446.

MR. MORGAN'S THEORY

As to Mr. Morgan's theory, in his Ancient Society (1877), of a movement of sanitary and moral reform, which led to prohibition of 'consanguine marriages' I shall return to it in a later part of this essay ('Other Bars to Marriage'). Here it will be found that Mr. Morgan is the source of certain other theories which we are to discuss, a fact involving a certain amount of repetition of arguments already advanced.

RETURN TO THE AUTHOR'S THEORY

We conclude, provisionally, that exogamy, for various reasons of sexual jealousy, and perhaps of sexual superstition, and of sexual indifference to persons familiar from infancy, may, at least, have tended to arise while each little human group was anonymous; before the acceptance of totem names by local groups. But this exogamous tendency, if it existed, must have been immensely reinforced and sweepingly defined when the hitherto anonymous groups, coming to be known by totem names, evolved the totem superstitions and tabus. Under these, I suggest, exogamy became fully developed. Marriage was forbidden, amours were forbidden (there are exceptional cases), within the totem name. This law barred, of course, marital relations between son and mother, between brother and sister, but, just as it stood, permitted incest between father and daughter, so long as the totem name was inherited from the mother. But that form of incest, in turn, came to be barred by another set of savage rules, which, whatever their origin, prohibit marriage within the generation. That set of rules, noted specially in Australia and North America, is part of what is usually styled 'The Class System.'

CHAPTER II

THE CLASS SYSTEM

Under this name appear to be blended, (1) the prohibition to marry within a division, which, in its simplest form, is said to cut the tribe into two 'classes' or 'phratries,' or 'groups;'[61] (2) the prohibition to marry within the totem name; (3) the prohibition to marry within the generation, and within certain recognised degrees ('classes,' 'sections') of real or inferred kinship—'too near flesh,' too close consanguinity, which, in their present condition, many Australian tribes undoubtedly regard as a bar to matrimony. But it does not follow that they originally held this opinion.

We shall first examine what authorities who differ from me, call the great 'bisection' of the tribe, into, say, Matthurie and Kirarawa, members of which must intermarry, the totem prohibition also remaining in force. It will here be suggested, in accordance with what has already been said, but contrary to general opinion, that the totemic prohibition is earlier than the prohibition of marriage between persons of the same segment of the 'bisection.' The opinions of most students appear, at present, to be divided thus. We hear that:

1. The exogamous division into two moieties, or 'phratries,' is earlier than the division of each into numerous totem kins. The totem kins are regarded as later 'subdivisions' of, or additions to, the two 'original' moieties.

2. Totem groups are earlier than the 'bisection' (though somehow, according to the same authors, the two moieties of the bisection bore totem names), but, before the 'bisection,' these totem groups were not exogamous. They only became exogamous when six of them, say, were arranged in one of the two moieties (phratries), now forbidden to marry, and another six in the other.

I venture to prefer, as already indicated, the system (3) that totem groups not only existed, but were already exogamous, before the great 'bisection' producing the 'phratries' came into existence, though I argue that 'bisection' is a misleading term, and that the apparent division was really the result of an amalgamation of two separate and independent local totem groups.

---

[61] Apparently, among the Kamilaroi, members of the same phratry may intermarry, avoiding unions in their own totems. Mathews (Proc. Roy. Soc. N.S.W. xxxi. 161, 162). Mr. Mathews calls a 'phratry' a 'group.'

This theory (presently to be more fully set forth) is original on my part, at least as far as my supraliminal consciousness is concerned. I mean that I conceived myself to have hit on the idea in July 1902. But something very like my notion (I later discovered) had been printed by Dr. Durkheim, and something not unlike it was propounded by Herr Cunow (1894). Mr. Daniel McLennan had also suggested it: and I find that the Rev. John Mathew had stated a form of it in his Eagle-Hawk and Crow (1899), (pp. 1922, 93-112). Mr. Mathew's hypothesis, however, involves a theory of contending and alien races in Australia. This theory does not seem well based, but, however that may be, I recognise that Mr. Mathew's hypothesis of the origin of exogamy (p. 98), and of the origin of the 'phratries' or 'primary classes,' in many respects anticipates my own. He opposes Mr. Howitt's conclusions, and I may be allowed to say that I would prefer Mr. Howitt, owing to his unrivalled knowledge, as an ally. On the other hand, the undesigned coincidence of Dr. Durkheim's, Mr. Daniel McLennan's, Mr. Mathew's, and Herr Cunow's ideas with my own, raises a presumption that mine may not be untenable.

THE CLASS SYSTEM IN AUSTRALIA

Though the existence of what are called exogamous 'phratries' (two to each tribe) was made known, as regards the North American tribes, by Mr. Lewis Morgan (to whose work we return) in the middle of the nineteenth century, almost our earliest hint of its existence in Australia came from the Rev. W. Ridley, a learned missionary, in 1853-55. In Mr. McLennan's Studies in Ancient History[62] will be found an account of Mr. Ridley's facts, as they gradually swelled in volume, altered in character, and were added to, and critically constructed, by the Rev. Mr. Fison, and Mr. A. W. Howitt. These gentlemen were regarded by Mr. McLennan as the allies of Mr. Morgan, in a controversy then being waged with some acerbity. He, therefore, criticised the evidence from Australia rather keenly. It is probable that Mr. Morgan and Mr. McLennan both had some right on their parts—seeing each a different side of the shield—though a few points in the discussion are still undecided. But it seems certain that the continued researches of Messrs. Fison and Howitt, reinforced by the studies of Messrs. Spencer and Gillen in Central Australia, have invalidated some of Mr. McLennan's opinions as to matters of fact.

---

[62] Second series, pp. 289-310.

Much trouble and confusion will be saved if we remember that, as has been said, under the 'classificatory system,' three sets of rules applying to marriage exist. The totem rule exists, rules as to marriage in relation to generations and so-called degrees of kindred (real or 'tribal') exist ('classes'), and, thirdly, there are the rules relative to 'phratries,' the phratries, being, I think, in origin themselves totemic. We shall mainly consider here the so-called 'bisection' of a tribe into two exogamous and intermarrying 'phratries,' while remembering Herr Cunow's opinion that a 'class' is one thing, a 'phratry' quite another.[63]

THE VARIETIES OF MARRIAGE DIVISIONS IN AUSTRALIA

Perhaps the most recent, lucid, and well-informed writer on the various divisions which regulate the marriages of the Australian tribes is Mr. R. H. Mathews.[64] In some regions, the system of two intermarrying phratries exists, without further subdivision (except in regard to totem kins). Sometimes each phratry is divided into two 'sections' (or 'classes'), making four for the tribe. Again, each phratry may have four 'subsections' or 'classes,' making eight for the tribe. Each phratry, like each 'class,' 'has an independent name by which its members are easily recognised.'

Obviously we need, of all tilings, to know the actual meanings of these names, but we do not usually know them. As we shall see, where a tribe has two 'phratries' and no subordinate 'classes,' the names of these 'phratries,' when they can be translated, are usually names of animals. In a few cases, as will later appear, when there are 'classes' under and in the 'phratries' their names seem to indicate distinctions of 'old' and 'young.' But Mr. Mathews nowhere, as far as I have studied him, gives the meanings of the 'class' names, some of which are of recent adoption. Mr. Mathews usually gives only 'Phratry A' and 'Phratry B.' We now cite his tables of the simple

---

[63] I shall, for my own part, use 'phratry' for the two 'primary exogamous divisions' of a tribe, and 'class' for the divisions within the 'phratry' which do not appear to be of totemic origin. Mr. Fison applies 'class' to both the primary divisions and those contained in each of them, observing that 'the Greek "phratria" would be the most correct term.' He is aware, of course, that this employment of phratria is arbitrary, but it is convenient. While he applies 'class' both to 'the primary divisions of a community, and their first subdivisions,' to the latter I restrict 'classes,' using phratry for the former (Kamilaroi and Kurnai, p. 24).

[64] Jour. and Proc. of the Roy. Soc. N.S.W., xxviii, xxxii, xxxiv.

'phratry' system, of the 'phratry' plus two classes system, and of the 'phratry' plus four classes system; making four, or eight, such divisions for the tribe.

'In describing the social structure of a native Australian community, the first matter calling for attention is the classification of the people into two primary divisions, called phratries, or groups—the men of each phratry intermarrying with the women of the opposite one, in accordance with prescribed laws.'

Mr. Mathews then mentions that some tribes have (1) this simple division only (of course, as a rule, plus totem kins). (2) Elsewhere each phratry is composed of two 'sections' (called by us 'classes'). (3) Elsewhere, again, each phratry has four sections (we need not discuss here the tribes where none of these things exist).

Mr. Mathews now gives tables representing the working of the system in each of the three cases.[65]

1

	Father	Mother	Son	Daughter
Phratry A	Kirraroo	Matturrin	Matturri	Matturrin
Phratry B	Matturri	Kirrarooan	Kirraroo	Kirrarooan

2

	Father	Mother	Son	Daughter
	{Murri	Buta	Ippai	Ippatha
Phratry A	{Kubbi	Ippatha	Kumbo	Butha
Phratry B	{Kumbo	Matha	Kubbi	Kubbithai
	{Ippai	Kubbitha	Murri	Matha

3

	Father	Mother	Son	Daughter
	{Choolum	Ningulum	Palyarin	Palyareenya

---

[65] Proc. Roy. Soc. N.S.W. xxxiv. 120-122.

Phratry A				
{Cheenum	Nooralum	Bungarin	Bungareenya
{Jamerum	Palyareenya	Chooralum	Nooralum
{Yacomary	Bungareeny	Chingulum	Ningulum

	Father	Mother	Son	Daughter
	{Chingalum	Noolum	Yacomary	Yacomareenya
Phratry B	{Chooralum	Neenum	Jamerum	Neomarum
	{Bungarin	Yacomareenya	Cheenum	Neenum
	{Palyarin	Neomarum	Choolum	Noolum

It will be seen that, under the simple phratry system, children of the female Matturrin are always Matturri and Matturrin, children of the female Kirrarooan are always Kirraroo and Kirrarooan. On the phratry plus two classes system, female Butha is mother of Ippatha and Ippatha of Butha for ever. On the phratry plus four classes system, female Ningulum has a Palyareena daughter, who has a Nooralum daughter, who has a Bungareenya daughter, whose daughter reverts to the original Ningulum class, and so on, ad infinitum. The women remain constant to their 'phratry,' and marry always the men of the opposite phratry.

It is to be observed that, by customary law, brothers and sisters actual (and not 'tribal') may never intermarry.[66] In short, consanguinity is now fully understood by the natives, and too close unions are forbidden on the ground of consanguinity. It also seems that, though the blacks are all on the same level of material culture, yet reflection on marriage rules, and modification of these rules by additional restrictions and alterations, have been carried much further by some tribes than by others. I by no means deny, but rather affirm, that consanguinity is now understood, and that rules have in some tribes been consciously made, and altered, to avoid certain marriages as of 'too near flesh.' But I do not think that, at the beginning, the objection to consanguineous marriages, as such, can have been entertained, and I am not of opinion that, for the purpose of preventing such marriages, in the beginning, a horde was bisected into two phratries, and each phratry split up into totem groups. Rather, I conceive, certain primitive conditions of life led to

---

[66] Prov. Jour. Roy. Soc. N.S.W., xxxiv. 127. Mr. Fison makes an exception for some Kamilaroi.

the evolution of certain rules, independent of any theory about the noxiousness or immorality of marriages of near kin; and then reflection on those primal rules helped to beget moral ideas, and improvements on the rules themselves. In the original restrictions, morality, in our sense, was only implicitly or potentially present, though now it has risen into explicit consciousness. The tribes came to think certain marriages morally wrong, or physically noxious, because they were forbidden; such unions were not, in the first instance, forbidden because they were deemed physically injurious, or morally wrong. These ideas have, by this time, been evolved; but it does not follow that they were present at the beginning.

I took the liberty of laying a brief sketch of my own theory before Mr. Howitt, who, after considering it, was unable to accept it. He was kind enough to send me a summary account of many varieties of institutions, which, as we have seen, prevail—from tribes with totems and the simple phratry and female descent, up to tribes which have lost their classes and totems, count descent in the male line, and permit marriage only between persons dwelling in certain localities, or not of 'too near flesh.' All sorts of varieties of custom, in fact, prevail. Again, the most backward tribes, in Mr. Howitt's opinion, have group-marriage;[67] the more advanced have individual marriage, with rare reversions on special occasions. Each advance, from mere phratry to phratry plus eight 'classes,' reduced the number of persons who might intermarry, and extended the range of exogamy (except where, as among the Arunta, the totem prohibition has ceased to exist). The marked tendency of the developing rules is to prevent marriage between persons 'too near in flesh,' or 'of the same flesh.' Mr. Howitt argues that, if the later stages of prohibition are the result of deliberate intention to prevent too near marriage, we may infer that the original 'bisection' of the 'undivided commune' was also consciously designed to prevent unions of persons of too near flesh.

To this I would reply, that the circumstances were different. The savages of recent centuries have been trained in the totem and phratry systems, and have now, like Mr. Howitt, excogitated the theory that these were originally designed for the purpose of preventing marriages of 'too near flesh,' wherefore all such marriages (even if permitted by the totem law) must be morally or materially evil. This is the theory expressed in the myths of the Dieri, Woeworung, and others; and it is the theory of many scientific writers. In brief, it is the hypothesis of men already trained to think near marriages morally wrong, or physically

---

[67] This view is discussed later.

injurious. But how could this idea occur to members of 'an undivided commune,' who had never known anything better?

That is the difficulty; and we get rid of it by disbelieving in a primeval undivided commune; and by supposing a long past of forbidden unions, the prohibition then resting on no moral ideas, but on the interest of the strongest, the jealousy of the adult sire. These prohibitions later evolved into conscious morality; and were at last susceptible of improvement by deliberate design. I shall now examine more in detail the ideas which do not win my assent.

MR. FISON ON THE GREAT BISECTION

In 1880, in Kamilaroi and Kurnai,[68] Mr. Fison, a learned missionary and anthropologist, gave his account of the organisation of certain Australian tribes. He speaks of (1) The division of a tribe, or community, into two exogamous intermarrying classes.[69] (2) 'The subdivision' (mark the phrase) 'of these classes into four.' (3) 'Their subdivision into gentes, distinguished by totems, which are generally, though not invariably, the names of animals.'

Now totems we know, and we have cited Mr. Mathews for the other divisions. Take (1) 'the two exogamous intermarrying classes.' Examples are

Male, Kumite; female, Kumitegor (one 'class,' which I call 'phratry').
Male, Kroki; female, Krokigor (the other 'class,' 'phratry').

Again.

Male, Yungaru (opossum); female, Yungaruan.
Male, Wutaru (kangaroo); female, Wutaruan.

What are these two 'primary' exogamous divisions? And why call them 'primary'?

'PRIMARY CLASSES?'

My object, as has been said, is now, contrary to general opinion, to repeat that the great dichotomous 'division' of a tribe into two exogamous, intermarrying, 'classes' or 'phratries,' is not

---

[68] P. 27 et seq.
[69] There is a tradition of an aboriginal Adam, who had two wives, Kilpara and Mukwara, these being the names of two phratries. On this showing brothers married paternal half-sisters (Kamilaroi and Kurnai, p. 33).

'primary' at all, but is secondary to groups at once totemic and exogamous, and is not, in origin, a bisection, but a combination. If I am right, the consequences will be of some curiosity. First, it will appear that the 'primary divisions' are themselves totemic in origin, thus implying the pre-existence of Totemism. Next it will be made to appear probable that the pre-existing totems were already exogamous before the phratries arose, and that exogamy does not date, as the best authorities hold, from the making of the great dichotomous divisions or 'phratries.' For no such dichotomous division, I suggest, was ever made.

THE 'PRIMARY DIVISIONS' ARE THEMSELVES TOTEMIC AND EXOGAMOUS

We see that, of the two 'phratries' Yungaru and Wutaru, Yungaru is 'opossum' (according to Mr. Chatfield) or 'alligator' (according to Mr. Bridgman); while Wutaru is 'kangaroo.' These two primary 'phratries,' therefore, have totemic names, and (in my opinion) were originally two local totem groups, each containing members of various totems derived from alien mothers. The same thing may be true when the meanings of the 'primary class names' ('phratries') can no longer be discovered. If so, the 'primary divisions' are, in origin, mere totem distinctions, involving, I think, the pre-existence of the rule of exogamy, which is also involved in the rules of the 'primary divisions.' Mr. Fison writes (what is obvious) 'in some places the primary divisions are distinguished by totem names at the present day.'[70]

'Probably they were so distinguished everywhere, in ancient times,' he adds, and this is certainly the case in North America, as we shall see later. Mr. Fison's opinion is my own so far, and, if it is right, if the 'primary class divisions' ('phratries'), within which marriage is now forbidden, were originally two totem divisions, then Totemism is earlier than the 'primary divisions.' On this point Messrs. Fison and Howitt say that the divisions on which marriage regulations are based 'are denoted by class names or by totems—frequently by both class names and totems.' In a note they add, 'Class names, so called by us solely for the sake of convenience, and because they cannot always be positively asserted to be totems, though the strong probability is that they are always totems.'[71]

By 'class names' the authors, I think, here mean the names of

---

[70] Kamilaroi and Kurnai, p. 40.
[71] J. A. I. xiv. 142.

the 'primary exogamous divisions' or 'phratries.' These are often, if not always, known by totem names. But the 'classes,' as distinguished from the 'phratries,' are not known by totemic names, as far as I am aware. Herr Cunow, we shall see, asserts that in some cases they denote mere seniority, 'big' and 'little,' 'young' and 'old.' Unless they can be proved to be totemic, we must, I repeat, carefully avoid confusing the 'classes,' four or eight, with the 'phratries,' in which they are included. The confusion is general and very misleading.

Totemism, according to Mr. McLennan, preceded exogamy, and made exogamy possible. Thus totem distinctions, with exogamy, may be older than the 'two primary class exogamous divisions,' in which, according to most authorities, exogamy began. Mr. Tylor is cautious: 'the dual form of exogamy' (the 'phratries,' or 'two primary divisions') 'may be the original form,' or at least that view is tenable.[72] The origin of exogamy is, however, unknown, in Mr. Tylor's opinion, which commits him to nothing.

Mr. Howitt, if I do not misinterpret him, also regards the two divisions, 'phratries,' as primary, but at the same time agrees with me, and Mr. Fison, that the two 'phratry' divisions were themselves in origin totemic.

THE TOTEM DIFFICULTY

At this point I lose Messrs. Fison and Howitt. I do not know what they mean, and, unless I misconstrue them, they unconsciously hold different opinions at different moments. They start with an 'undivided commune.' Mr. Fison, however, is not certain on this point. To prevent near marriages (previously universal), the commune is split into two exogamous intermarrying phratries. The names of these phratries are totemic, and each phratry has its totem. Such is their theory. How and why?

Did totemic divisions already exist in 'the undivided commune'? If so, the commune was not undivided! Or were totem names given, nobody knows why, to the two phratries at the time when the 'bisection' of the commune was made? Did the legislator send half the horde to the right, crying, 'You are sheep,' and half to the left, saying, 'You are goats,'—or rather, say, Emus and Kangaroos? This is not easily thinkable. But, if this was done, whence came the other totem kins, often numerous, within each phratry?

---

[72] Journal of the Anthropological Institute, xviii. 264.

Mr. Fison says that the totem kins (or 'gentes') 'arose out of two primary divisions, by an orderly process of evolution, such as might be expected from the forces at work,' and 'we have seen how' the phratries subdivided 'into other subdivisions, distinguished by totems.'[73] But, alas, I have seen nothing of the sort! Mr. Fison has merely asserted the fact. 'The totems affect the intersexual regulations ... by narrowing the range of matrimonial selection.'[74] Here would be a reason for the evolution of these totem kins. But this added restriction is exactly what (given phratries) the totems do not effect. There are so many totems in each phratry, but as the same totem (except among the Arunta and similarly disorganised tribes) never occurs in both phratries, the range of sexual selection is thus not more restricted by the totem than by the phratry. The members of each phratry may not intermarry, and all persons of their totem are in their phratry and so are not marriageable to them. They would all be exactly as exogamous as they are, if there were no totem rules, nothing but phratry rules. Thus the totems cannot be later deliberate segmentations of the phratry, for additional exogamous purposes, because they serve no such purpose, except where, among the Kamilaroi, a man may marry in his phratry, if he marries out of his totem. But that is a peculiarity.

Mr. Mathews writes, 'Under the group' (phratry) 'laws it is impossible for a Dilbi or Kupathin' (phratry names of the Kamilaroi) 'to marry a woman bearing the same totem name as himself, for the reason that such a totem does not exist in the division' (phratry) 'from which he is bound to select his wife. But when persons of the same group' (phratry) 'were permitted to marry each other, it became necessary to promulgate a law prohibiting marriage between persons of the same totem.'[75] But there were totems before that novelty of marriage within the phratry, and why were they there? Moreover, under phratry laws it was already the rule that no man could marry a woman of his own totem. Obviously we are not told how the totem kins arose out of the phratries, 'by an orderly process of evolution such as might be expected from the forces at work.' One sees no reason at all for the rise of totem kins within the phratry, itself, by Mr. Fison's theory, originally totemic.

Totem kins are called 'subdivisions' by Mr. Howitt, but why were the phratries subdivided into totem kins, and why were there totem groups in 'the undivided commune' before the bisection, the phratries (the result of the bisection) being themselves, in Mr.

---

[73] Kamilaroi and Kurnai, p. 107.
[74] Op. cit. p. 41.
[75] Proc. Roy. Soc. N.S.W. xxxi. 162.

Howitt's hypothesis, totem groups? I quote a statement of the case by Mr. Howitt (1889): 'The fundamental principle of aboriginal society in Australia is the division of the community into two exogamous intermarrying moieties. Out of this division into two groups, and out of the relations thus created between the contemporary members of them and their descendants, the terms of relationship must have grown. As the two primary divisions (classes)' ('phratries') 'have become again divided in the process of social development, and as the groups of numerous totems have been added,' &c.[76]

Here the totem kins are not orderly evolved out of the phratries, nor subdivided out of them, but are 'added.' Where were they picked up, whence did they arise, why were they 'added'?

May we not conclude that no clear account, or theory, of the origin and purpose of totems and totem kins has been laid before us?

Mr. Howitt elsewhere writes, 'If the supposition is correct that, in the primary divisions, we may recognise the oldest forms, and in the subdivisions somewhat newer forms of Totemism' (newer names of totems?), 'it should be found that these earlier divisions show signs of antiquity as compared to the totems which are, according to this hypothesis, the nearest to the present time. This, I think, is the case.' Thus, in fact, some of the Australian names for the two divisions are no longer to be translated,[77] perhaps owing to their antiquity, and sometimes the names are lost, as, elsewhere, in Banks Island. When translatable, the phratry names are totemic.

But this hardly amounts to proof that the 'primary divisions' are really older than totemic divisions, plus exogamy. The existing names of the 'primary divisions' may be older than existing totem names, in some cases. But that may be because the two 'primary divisions' endure, unchanged, while a local totem group may become extinct.[78] Its place, perhaps, may be filled up by a totem group of relatively recent name, or, perhaps, in a great trek into a land of novel fauna and flora, old totem names might be exchanged for new ones. 'Munki' (sheep) is said to have been recently adopted.[79] Mr. Fison here corroborates my suggestion. 'If a tribe migrate to a country in which their totem is not found, they will, in

---

[76] On the Organisation of Australian Tribes, p. 129; Transactions of Royal Society of Victoria, 1889.

[77] The natives retain sacred songs to Daramulun, but cannot (or will not?) translate them. Proc. Roy. Soc. N.S.W. xxxiv. 280.

[78] Spencer and Gillen, p. 152.

[79] Howitt, J. A. I. xviii. 37-39.

all probability, take as their totem some other animal which is a native of the place.'[80]

Mr. Howitt, then, believes that 'the primary class divisions' were originally totemic, and also that the 'class system' as a rule has been developed through the subdivision of the earlier and simpler forms by 'deliberate arrangement.'[81]

This appears to mean that savages began by making two divisions, bearing totem names, and established them as primary exogamous divisions. Later they cut them up into slices, each slice with a newer totem name. Or the totem divisions are evolved within the phratry, somehow or other, as in one of Mr. Fison's views. Or they are 'added'—for what purpose? Thus every tribesman has now a 'class name' (phratry name)—an old totem name (say either Eagle-Hawk or Crow), and no Crow may marry an Eagle-Hawk. But, later, they split Crows up into, say, bats, rats, cats, and kangaroos, while they split Eagle-Hawk up into, say, grubs, emus, mice, and frogs. Now each person, under this arrangement, has two totem names. He is Eagle-Hawk (old) and (new) grub, emu, mouse, or frog: or he is Crow (old) and (new) bat, cat, rat, or kangaroo. If cat, he may not only not marry a Crow, but also he may not marry a cat. What could be the reason for this new subdivision of Eagle-Hawk and Crow, and for this multiplication of marriage prohibitions, which, given the phratries, prohibit nothing?[82] I shall try to show, and have already suggested, that, from a period infinitely remote, each member of the Eagle-Hawk and Crow local groups may also have been, or rather must have been, a grub, emu, mouse, or frog, bat, rat, cat, or kangaroo, by inheritance and birth. So understood, the 'primary divisions' (Eagle-Hawk and Crow) were not deliberately subdivided (as I conceive them to have been on Mr. Ho wit Vs system) into the other numerous new totem groups, nor were the totem kins added to the phratries, nor were they orderly evolved out of the phratries, but, from the dawn of Totemism with exogamy, they contained these totem groups within themselves; a fact which early man came to perceive.

Mr. Howitt adds, 'If the two first intermarrying groups' ('phratries') 'had distinguished names, they were probably those of animals, and their totems, and, if so, the origin of Totemism would be so far back in the mist of ages, as to be beyond my vision.' In the chapter on the 'Origin of Totemism,' we try to penetrate 'the mist of

---

[80] Kamilaroi and Kurnai, p. 235, note.
[81] Op. cit. pp. 59, 62, 63, 66.
[82] New marriage prohibitions may have been, and, I believe, were added, but the divisions thus made were not, I think, totemistic.

ages,' and to see beyond the range of vision of Mr. Howitt. But the 'Origin of Totemism' cannot be beyond Mr. Howitt's range of vision, if he agrees with Mr. Fison that the totem kins were orderly evolved within the phratry, or were segmented out of the phratry, or split off, as colonies, from the phratry (Dr. Durkheim's theory), or were added to the phratry, for some reason.

It seems, then, that he does not commit himself to any of these four theories. He appears to confess to having no theory of the origin of Totemism, which, in his opinion, gave the names to the phratries, these being the result of the primary bisection. Probably his best plan would be to say 'the horde was bisected into two moieties, for exogamous purposes, and animal names, for the sake of distinction, were arbitrarily imposed on the phratry divisions.' But, then, what about the many totem kins within the phratry? We receive no solid theory about them. They were certainly not arbitrarily marked out later, within the phratry, for exogamous purposes which they do not fulfil. If they were picked up elsewhere, and added into the phratry, where did they come from? Crowds of totems were not going about, Mr. Howitt seems to think, before the bisection, because, if so, we saw hordes were not 'undivided,' before the bisection, but were already divided into totem kins.

Or shall we say that the undivided communes had already organised distinct co-operative magical totem groups, to do magic for the good of the food supply, plants and animals, but that these totem groups were not exogamous before the bisection? After the bisection two of these magical totem groups, say Eagle-Hawk and Crow, were selected, shall we guess, to give names to the two moieties or phratries? The other totem groups fell, or were meted out, some into Crow, some into Eagle-Hawk. This is a thinkable hypothesis, but it is fatal to the theory of subdivision, or of segmentation, or of evolution, as causes of totem kins within the phratries; and it is not suggested by Messrs. Fison and Howitt.

Thus we must construct for ourselves, later, a theory of the Origin of Totemism. We are actually constrained to make this effort, because it will probably be admitted that, having no theory, or hesitating between three or four theories, of the origin of totems and of totem kins, Messrs. Fison and Howitt produce an hypothesis of the evolution of Australian society which cannot be construed by us into an intelligible form.

Mr. Howitt elsewhere writes, 'The existence of the two exogamous intermarrying groups' ('phratries') 'seems to me almost to require the previous existence of an undivided commune, from the segmentation of which they arose.'[83] But they, the phratries, were totemic, and why? Once again, why was the undivided commune divided? We know not the motive for, much less the means of effecting, such a great change 'in the beginning.'

In 1885, Messrs. Howitt and Fison were aware of, and expressed their sense of this difficulty (that of dividing people out into arbitrary groups) in the case of ancient Attica. Speaking of the γένος, or clan, in Attica, they combat the opinion of Harpocration, that the people were 'arbitrarily drafted into the γένη.[84] Our authors remark, 'Ancient society—the more ancient—does not thus regulate itself. Nascitur non fit. One can understand a Kleisthenes redistributing into demes a civilised community which has grown into a State, but the notion of any such arbitrary distribution of men into γένη; in the beginning of things cannot be entertained for a moment.'[85]

This being so, how can our authors maintain that, 'in the beginning of things,' given an 'undivided commune,' all its members were 'drafted' into one or other of two divisions, and again into totem groups. A subdivision of the 'phratries' into totem groups, by deliberate arrangement, is clearly as artificial and arbitrary as the scheme suggested by Harpocration, 'which cannot be entertained for a moment.'

We are speaking of 'the beginning of things,' not of the present state of things, in which we know that modifications of the rules, e.g. the division into eight 'classes,' are being deliberately adopted.[86] In 'the beginning of things,' as Messrs. Howitt and Fison, in 1885, maintained, society nascitur non fit. Our effort is to show the process of the birth of society before conscious and deliberate modifications were made to prevent marriages, of 'too near flesh.' Our criticism of Messrs. Fison and Howitt's theories may perhaps indicate that they are insufficient, or but dubiously intelligible. Something clear and consistent is required.

---

[83] Organisation of Australian Tribes, p. 136.
[84] Harpocration s.v. γεννῆται Greek: genneitai.
[85] J. A. I. xiv. 160.
[86] Spencer and Gillen, pp. 72, 420.

CHAPTER III

TOTEMS WITHIN THE PHRATRIES

AMERICAN SUPPORT OF THE AUTHOR'S HYPOTHESIS

The system which I advocate here, as to the smallness of the original human groups, and their later combination into larger unions, seems to have, as regards America, the support of the late Major Powell, the Director of the Bureau of Ethnology, and of Mr. McGee of the same department. This gentleman writes, 'Two postulates concerning primitive society, adopted by various ethnologic students of other countries, have been erroneously applied to the American aborigines ... The first postulate is that primitive men were originally assembled in chaotic hordes, and that organised society was developed out of the chaotic mass by the segregation of groups ...' This appears to be Mr. Hewitt's doctrine. In fact, Mr. McGee says, American research points, not to a primal horde, 'bisected' and 'subdivided' into an organised community, but to an early condition 'directly antithetic to the postulated horde, in which the scant population was segregated in small discrete bodies, probably family groups....' The process of advance was one of 'progressive combination rather than of continued differentiation.... It would appear that the original definitely organised groups occasionally coalesced with other groups, both simple and compound, whereby they were elaborated in structure....' Mr. McGee adds, 'always with some loss in definiteness and permanence.' As far as concerns Australia, I do not feel sure that the last remark applies, but, on the whole, Mr. McGee's observations, couched in abstract terms, appear to fit what I have written, in concrete terms, about the probable evolution of Australian tribal society.[87]

The theory thus suggested makes little demand on deliberate legislation, as we shall see later.

---

[87] Ethnological Bureau, Annual Report, 1893-1894, pp. 200, 201.

DELIBERATE ARRANGEMENT

This I take to be important. It seems well to avoid, as far as possible, the hypothesis of deliberate legislation in times primeval, involving so sweeping a change as the legal establishment of exogamy through a decree based on common consent by an exogamous 'Bisection' consciously made. Exogamy must have been gradually evolved. But, if we begin with Mr. Howitt's original undivided commune, and suppose a deliberate bisection of it into two exogamous phratries, each somehow containing different totems; or if we suppose a tribe of only two totems, and imagine that the tribe deliberately made these totems exogamous, which they had not been before, and then subdivided them into many other totem groups, we see, indeed, why persons of the same totem may not intermarry. They now, after the decree, belong to the same exogamous 'phratry' within which marriage is deliberately forbidden. But, on this theory, I find no escape from the conclusion that the 'bisection' into 'phratries' was the result of a deliberate decree, intended to produce exogamy—for the bisection has not, and apparently cannot have, any other effect. Now I can neither imagine a motive for such a decree, nor any mode, in such early times, of procuring for it common consent. At this point we have laboured, and to it we shall return, observing that our hypothesis makes much less appeal to such early and deliberate legislation.

TOTEMS ALL THE WAY

In any case, by Mr. Fison's and Mr. Howitt's theory and our own, we have totems almost all the way: totems in the so-called 'primary divisions' (phratries); totems in the so-called gentes, and all these divisions (setting the Arunta apart) are strictly exogamous. The four or eight 'classes,' on the other hand, are apparently not of totemic origin. However much the systems may be complicated and inter-twisted, the basis of the whole, except of the four or eight 'classes,' is, I think, the totem exogamous prohibition. There are many examples of the type; thus the Urabunna 'are divided into two exogamous intermarrying classes, which are respectively called Matthurie and Kirarawa, and the members of these again are divided into a series of totemic groups, for which the native name is Thunthunnie. A Matthurie man must marry a Kirarawa women' (as in the system of the Kamil-speaking tribes, or Kamilaroi, reported on by Mr. Fison)—'and not only this, but a man of one totem must marry a woman of another totem.' This is precisely what I should expect. It works out thus:

{ Old Local Totem Group }   Matthurie.
{ New 'Phratry' }
{ Old Local Totem Group }   Kirarawa.
{ New 'Phratry' }

Each of these 'phratries' has five totems, not found in the other class, and how this occurred, if not by actual deliberate arrangement, I do not know. One thing is clear: totem and phratry are prior to 'class' divisions. They occur where 'class' divisions do not. But my theory does not involve the deliberate introduction of exogamy, by an exogamous bisection of groups not hitherto exogamous, or by making two pre-existing totem groups exogamous. I take the groups to have been exogamous already, before the blending in connubium of two local totem groups (now 'phratries'), each including numbers of already exogamous totem kindreds. They were exogamous before the 'phratries' existed, and

DISTRIBUTION OF TOTEMS IN THE 'PHRATRIES'

Mr. McLennan, ere he had the information now before us, wrote, in 1865, 'Most probably contiguous groups would be composed of exactly the same stocks' (we can now, for 'stocks,' read 'totem kins')—'would contain gentes of precisely the same names.'[88] This is obvious, for Emu, Kangaroo, Wild Duck, Opossum, Snake, and Lizard, living in the same region, would raid each other (by the hypothesis) for wives, and each foreign wife would bring her own totem name into each group. Yet we find that the two 'primary classes' (phratries) of the Urabunna (which, on my theory, represent two primitive totem local groups, say Emu and Kangaroo, each with its representatives of all other totem groups within raiding distance) never contain the same totems.

It is mathematically impossible that this exclusiveness should be the result of accident. On a first consideration, therefore, I took it to be the result of deliberate legislative design, at the moment when on my hypothesis two local totem groups, containing members of several totems of descent, united in connubium. The totem names, I at first conceived, with reluctance, must have been consciously and

[88] Studies in Ancient History, p. 221.

deliberately meted out between the two local totem groups, now become phratries. This idea did not involve so stringent and useless a measure as that of segmenting the two phratries into minor totem groups: however the idea was still too much akin to that of Harpocration as regards the arbitrary drafting of the Attic population into γένη. But, on further reflection, I conceived that my first theory was superfluous. Given the existence of local groups, as such totemic, and of totem kins of descent within the original local totem groups, the actual results, I thought, arise automatically, as soon as two local totem groups agree to intermarry. Men and women must many out of their local totem group (now 'phratry') and must marry out of their totem of descent. Consequently, no one totem could possibly exist in both phratries. This I now, on third thoughts, 'which are a wiser first,' deem erroneous. The automatic arraying of one set of totems into one, or another set into the other, phratry, would not occur. The totems have been divided between the two phratries.[89] This condition of affairs is universal in

---

[89] Suppose we take a group ranging in a given locality, and known to its neighbours as the Emu group. Let us also take a similar and similarly situated Kangaroo group. Let us suppose that each such group has raided for its wives among Opossum, Grub, Cat, and Dingo groups. By female descent, both the Emu and Kangaroo groups will contain persons of the Opossum, Grub, Cat, and Dingo groups. This being so a man of the Emu local group, named Grub by totem, might marry a woman of the Emu local group, by totem of descent an Opossum; and similarly in the Kangaroo group. But, as Dr. Durkheim remarks in another case, 'the old prohibition', deeply rooted in manners and customs, survives (L'Année Sociologique, v. 107, note). Now 'the old prohibition' was that a man of the Emu group was not to marry a woman of the Emu group. That rule endures, though the Emu group now contains men and women of several distinct totem kins. To escape from the difficulty, by my theory, Emu local totem group makes connubium with Kangaroo local totem group. Any Emu man may marry any Kangaroo woman not of his own totem by descent. But this does not, automatically, throw Opossum and Grub into one, Cat and Dingo into another, of the two local totem groups, Emu and Kangaroo, now become phratries, with loss of their local character. For if a man, by phratry Emu, and by totem of descent Cat, marries a woman, by phratry Kangaroo, and by totem of descent Grub, their children, by female descent, are Kangaroo Grubs. Meanwhile, if a man, by phratry Kangaroo, and by totem Cat, marries a woman, by phratry Emu, and by totem Grub, their children are Emu Grubs. There are thus Grubs in both phratries, a thing that never occurs (except among the Arunta). Therefore the division of the totem kins, some into one phratry, others into the other, is not automatic. There might be a tendency, by way of making assurance doubly sure, for the totem kins to be assorted into the two phratries, but some kind of

45

Australia, except where, as among the Arunta and similar tribes, the same totem comes to exist in both phratries, so that men and women of the same totem, but of opposite phratries, may intermarry. That breach of old rule, we shall try to show, arises from the peculiar animistic philosophy of the Arunta, by virtue of which totems are no longer totems of descent, but are otherwise obtained. The Kamilaroi practice of interphratry marriage arises out of respect for totem and neglect of phratry law.

My conjecture takes for granted, let me repeat, that, before the 'bisection,' or the amalgamation, which produced the two exogamous 'classes,' the totem kindreds were already exogamous. My reasons for this opinion have already been given, in the discussion of Mr. Crawley's theory of the origin of exogamy (supra), to which the reader may refer. My suggestion makes the growth of exogamy non-moral, gradual, and almost unconscious, till it is clinched and stereotyped by the totem tabu.[90] The opposite theory—namely, the deliberate bisection into exogamous 'classes,' of totem groups, or of an 'undivided commune' not previously exogamous, appeals too much, I repeat, to conscious and—as far as we can see—motiveless legislation, at an early stage. The bisection must have had a purpose, and has no visible purpose except the establishment o f exogamy, and why did the 'undivided commune' establish that?

THE IDEAS OF MR. FRAZER HIS EARLIER THEORY

It cannot be concealed that my conjecture is opposed to the mass of learned opinion, which represents the primary 'phratries' as the first exogamous bodies, and the totems in each as later subdivisions of the phratries. The writers who, like Mr. Fison, recognise that the primary subdivisions are themselves, in origin, totem divisions, do not (as I understand) regard these very ancient totem groups as already exogamous, before the institution of 'phratries.'

Again, turning from Australia to North America, we find Mr. Frazer, at least in one passage, on the side of the view generally held. Of the 'phratry,' in America, he says, 'the evidence goes to

---

deliberate arrangement does seem necessary. The same necessity attends Dr. Durkheim's theory later criticised.

[90] See again Durkheim, in L'Année Sociologique, i. 47-57, on the superstition as to blood, and the totem as a sacred representative of the inviolable blood of the kindred. That superstition gives religious sanction to a pre-existing exogamous tendency.

show that in many cases it was originally a totem clan which has undergone subdivision.'[91] Many examples are then given of the North American 'phratries,' which include totem groups within them. 'The Choctaws were divided into two phratries, each of which included four clans' (totem kins); 'marriage was prohibited between members of the same phratry, but members of either phratry could marry into any clan of the other.' Among the Senecas, one phratry included the Bear, Wolf, Beaver, and Turtle totems: the other held the Deer, Snipe, Heron, and Hawk totems; just as in Australia. Among the Thlinkets and Mohegans, 'each phratry bears a name which is also the name of one of the clans' (totems) 'included in it;' Mr. Frazer adds, 'it seems probable that the names of the Raven and Wolf were the two original clans of the Thlinkets, which afterwards by subdivision became phratries.'[92] This is precisely as if we were to argue that Matthurie and Kirarawa were the 'two original clans' of the Urabunna, 'which afterwards by subdivision' (into totem groups) 'became phratries,' or 'primary exogamous divisions.'

The objections to this theory, as advocated by Australian inquirers, apply to the American cases as interpreted here by Mr. Frazer. In the first place, how are we to conceive of a large tribe, like the Thlinkets, as originally containing only two totems, Raven and Wolf?[93] If we do take this view, we seem almost driven to suppose that, in exceedingly early times, the Thlinkets deliberately bisected themselves, for some reason, called one moiety Ravens, the other moiety Wolves, and then made the divisions exogamous. Or, perhaps, having two totems and only two, Raven and Wolf, they deliberately decided that members of neither group should marry within itself; but should always take wives from the other group. Later, the two tribes, Raven and Wolf, again deliberately subdivided themselves, or perhaps, as in Dr. Durkheim's view, Wolf threw off colonies which became five totem kins, and Raven threw off colonies which became five other totem kins.

Is it not more readily credible that, over a large extent of Thlinket country, many small local groups came, by an unconscious process (see 'The Origin of Totemism'), to bear each a separate totem name? The two most important local groups, Raven and

---

[91] Totemism, p. 60 (1889).

[92] Totemism, p. 62.

[93] The people of New Britain group of islands are divided into two exogamous sets. The totems of these classes are two insects, but I incline to suppose that there are, or may have been, totem kins included within these totemic classes. Our informant, the Rev. B. Danks, regrets that he did not pay more attention to these matters. J. A. I. xviii. 281-294.

47

Wolf, would inevitably each contain, by the working of exogamy and female kin, members of all the other totems which would array themselves, five in each chief group, Raven and Wolf, as I have conjectured in speaking of the Australian cases.[94]

Again, I cannot believe that a tribe like the Thlinkets originally had but two totems, not yet exogamous, then made them exogamous, and then cut them up, or let them split off, into many exogamous totem groups. No motive is obvious: the people, by the theory, being exogamous already.

OBJECTIONS TO MR. FRAZER'S EARLY THEORY

We shall later see that Messrs. Spencer and Gillen appear to advance, but also to qualify out of existence, a theory of a motive for an exogamous bisection of earlier non-exogamous local totem groups. They practically explain away their own explanation of—the great bisection, but it rests, while it exists, on certain recently discovered facts, which, in turn, are fatal, perhaps, to any theory that a tribe had originally but two totems, which became 'phratries,' on being subdivided into other totems. The new facts accepted and theorised on by Mr. Frazer and Mr. Spencer, would make it seem perhaps impossible that a tribe like the Thlinkets should originally have possessed but two 'clans' or totems. The facts, as stated by Mr. Spencer, in 1899, are these, or rather, this is his hypothesis founded on his facts. 'In our Australian tribes the primary[95] function of a totem group is that of ensuring, by magic means, a supply of the object which gives its name to the totem group.'[96] Mr. Frazer says, 'in its origin Totemism was, on our theory, simply an organised and co-operative system of magic.... Each totem group was charged with the superintendence and control of the particular department of nature from which it took its name....'[97]

---

[94] On the other hand, among the Mohegans, I can admit that Little Turtle, Mud Turtle, and Great Turtle may be deliberate subdivisions of the Turtle totem, now a phratry, but even this need not necessarily be the case; the different species of turtles being quite capable of giving names to different totems. I would not deny the possibility of the occasional segmentation of a totem group—far from it—but I doubt whether great tribes originally (and, as it seems, deliberately) first bisected themselves, and then cut up the two main divisions.

[95] My italics.

[96] J. A. I., N.S. i. 278.

[97] Ibid. p. 282.

But this is hardly the origin of Totemism, so long as we are not told how, or why, each totem group took its name from a department of nature. Had it the name, before it worked magic for its eponymous object, or did it take the name because it worked the magic?

Again, there are dozens of such departments,[98] which implies the existence of dozens of organised and co-operative totem groups: not of an original poor pair of such groups alone. Can we believe that, on Mr. Frazer's earlier theory, the Thlinkets formed but two such groups, one 'charged with' the duty to mollify the Wolf, the other to take care of the interests of the Raven? Manifestly this is unlikely. I elsewhere oppose this theory of the magical Origin of Totemism, made at first to fit the case of the Arunta and cognate tribes. If organised co-operation in magic is the source of Totemism, we may be pretty confident that no tribe began by appointing one half of all its members to do magic to propagate ravens, and the other half to mollify wolves. This would indicate, in the magical and co-operative tribe, a most oddly limited and feebly capitalised flotation of the company—merely 'Wolf and Raven.' No tribe would select ravens as the article of food which most required careful propagation and preservation, even if the Wolf most demanded to be propitiated and mollified. The new Australian facts (whatever their interpretation) are fatal to the older idea that a tribe could have had only two original totems: an idea which we may perhaps regard as now abandoned, at least by Mr. Frazer.

Thus Mr. Spencer himself remarks that, in Arunta tradition, there were numbers of totem groups before the great dichotomous division was made. That is my own opinion: though I do not hold it for Mr. Spencer's reasons, or believe in any 'bisection.'

MR. SPENCER'S THEORIES OF THE BISECTION

It will be noted that Mr. Spencer's original totem groups existed for magical purposes only, and were not exogamous.

'The traditions of the Arunta tribe point to a very definite introduction of an exogamic system long after the totemic groups were fully developed, and, further, they point very clearly to the fact that the introduction was due to the deliberate action of certain ancestors. Our knowledge of the natives leads us to the opinion that

---

[98] Mr. Mathews counts thirty-four totems in the Dilbi, and as many in the Rupathin 'phratries.' Proc. Ray. Soc. N.S.W. xxxi. 157-158.

it is quite possible that this really took place, that the exogamic groups were deliberately introduced so as to regulate marital relations.'

The Arunta 'exogamic groups' are 'classes,' and 'phratries,' the totem does not now regulate marriage among the Arunta. I shall later try to show, that, originally, totems did regulate marriage, among the Arunta. But here we find Mr. Spencer averring that possibly 'the exogamic groups were deliberately introduced so as to regulate marital relations' among the Arunta. This opinion surprises us, if we hold that exogamy was, in its original forms, the result, not of a deliberate enactment, but of gradual and unconscious processes, to which, later, conscious modifications have been added. Mr. Spencer, despite the passage cited, is obviously of the same opinion, for he proceeds to remark, 'By this we do not mean that the regulations had anything whatever to do with the idea of incest, or of any harm accruing from the union of individuals who were regarded as too nearly related.... It can only be said that far back in the early history of mankind, there was felt the need of some form of organisation, and that this gradually resulted in the development of exogamous groups.'

This statement must remind us of what the ancient ballad sings about Lord Bateman:

> He shipped himself all aboard of a ship,
> Some foreign country for to see.

The scholiast (Thackeray, I think) explains, 'some foreign country he wished to see, and that was the extent of his desire: any foreign country would serve his purpose, all foreign countries were alike to him.' In the same way, long ago, the ancestors of the Australians 'felt the need of some form of organisation,' and that was the extent of their desire; any organisation would serve their purpose. Nevertheless, Mr. Spencer also says that, quite possibly, 'the exogamic groups were deliberately introduced so as to regulate marital relations.' But exogamic groups can regulate marital arrangements in one way only—that is, by introducing exogamy. Yet Mr. Spencer remarks that 'the development of exogamic groups' gradually resulted from some organisation of unknown nature. I am unable to reconcile Mr. Spencer's statements with each other. The 'bisection' of his theory could not, I fear, be 'gradual.'

Mr. Frazer, in 1899, begins with numerous totem groups, primarily and originally arranged for mere purposes of co-operative magic, in the social interests of a large friendly tribe, itself no primitive institution, one thinks. Then he supposes that the exogamous bisection occurred (and why did it occur?), and then 'if

50

the existing totem groups were arranged, as they naturally would be, some in one of the two new classes, and the rest in the other, the exogamy of the totem groups would follow, ipso facto.'[99] Mr. Frazer does not here pretend to guess why the bisection occurred. The rest is quite obvious: but it is unavoidably inconsistent with Mr. Frazer's earlier theory, that a tribe begins (or that the Thlinkets began) with two original totem groups, made them exogamous, and then 'subdivided' them up (or did they merely swarm off?) into many totem groups. It is against that almost universal theory, in 1899 abandoned (as I conceive) by Mr. Frazer, that I have so long been arguing. There was not first an exogamous bisection of a tribe, or the addition of the exogamous rule to two 'original clans,' or totem groups, and then the subdivision of each of the two sections into a number of totems. This cannot have occurred. Totems, I venture to think, did not come in that way, but pre-existing totem kins, granting the bisection, might fall into one or other phratry, if they had always been exogamous.

ADVANTAGES OF THE SYSTEM HERE PROPOSED

On my system, as has been already stated, the origin of exogamy may have been sexual jealousy, in small primitive groups, perhaps aided by 'sexual tabu,' with the strange superstitions on which it is based, and these causes would be strengthened enormously by the totem superstition, later. The totem name would now be the exogamous limit. The 'phratries' might result, quite naturally, and even gradually, now in one region, now in another, from the interlocking and alliance, with connubium, of two large friendly local totem groups, an arrangement of which the advantages are so obvious that it might spread by way of imitation and accretion.

This view of the possible origin of what is usually called the 'bisection' of 'the undivided commune' had already been suggested by the late Mr. Daniel McLennan.[100] Writing before our information was so full as it now is, he says, as to the two 'phratries' Kumite and Kroki (answering to Matthurie and Kirarawa), 'were it worth while to make surmises, it would not be unreasonable to surmise that at Mount Gambier two separate local tribes[101] containing different totem kindreds had, through the operation of exogamy and female

---

99 J. A. I., N.S. i. 284-285.
100 Studies in Ancient History, second series, p. 605.
101 Local totem groups, in my theory.

kinship, become welded into one community.' Mr. Daniel McLennan, unluckily, inherited his brother's feud against Mr. Fison, and he opposed all that gentleman's doings. Later research has corroborated many of Mr. Fison's facts, and extended the range of their influence. On this point, however—namely, that the 'phratries' are not the result of a bisection, but of an amalgamation—Mr. Daniel McLennan appears to have had a good case. He illustrates his theory, and mine, by remarks on a tradition of the tribes of Northern Victoria.[102]

The exogamous 'phratries' of these tribes are Eagle-Hawk and Crow. The tradition represents these birds as hostile creative powers. They made peace on the terms 'that the Murray blacks should be divided into two classes' ('phratries'), 'the Makquarra, or Eagle-Hawk, and the Kilparra, or Crow.... Out of the enmities' (of the original Crow and Eagle-Hawk) 'arose the two classes, and thence a law governing marriage among these classes.' This tradition, it will be observed, espouses the theory of a bisection, deliberately made of 'the Murray blacks,' into two intermarrying and exogamous classes. Mr. McLennan writes, 'But what the tradition suggests is, not that the Crow and Eagle agreed to divide one tribe into two, with a view to the better regulation of marriage, but that Crow and Eagle or Eagle-Hawk were tribes (and they might have been constituted in the ordinary Australian way) which long waged war against each other, and that at length there came peace, and then their complete interfusion by means of friendly marriages.' The tradition asserts the reverse; it adopts, or rather it forestalls, the scientific theory of a 'bisection' of the Murray blacks, not the amalgamation of two tribes (or large local totem groups). But I agree with Mr. McLennan in prefering, for the reasons given, the theory of an amalgamation. It is rather curious and interesting to observe that almost every scientific hypothesis about totems and 'classes,' which I am obliged to reject, has, in fact, been forestalled by the theories which the natives themselves express in their explanatory myths. Myths, I fear, are never in the right. 'The aborigines themselves,' says Mr. Howitt, 'recognise the former existence of the undivided commune in their legends, but,' he judiciously adds, 'I do not rely upon this as having the force of evidence.'[103]

We shall presently see that other distinguished anthropologists do, to some extent, rely on Arunta myths, as 'bearing the stamp of authenticity.' The truth is that the native

---

[102] Brough Smyth, Aborigines of Victoria, i. 423-424.
[103] On the Organisation of Australian Tribes, p. 186.

thinkers have hit on the same hypothesis as their European critics, the hypothesis of something like deliberate primeval legislation to a given end, the regulation of marriage. Far from accepting any such native myths, I am rather inclined to hold that, whatever theory be correct, the theory of the savage myth-makers must be wrong. It ought to be said that Mr. Fison, at least, knows what his own theory involves, and once even frankly accepted the possibility that the Dieri myth (the foundation of exogamy by divine decree) may be historically true. 'All I contend for is,' he says, 'that if the former existence of the undivided commune be taken for granted' (and Mr. Fison, unlike Mr. Howitt, regards the undivided commune as a mere unproved hypothesis), 'its division into exogamous clans must have had precisely the effect' (a consciously reformatory effect) 'which Mr. Morgan's theory requires. If such a community ever existed, I do not hesitate to say that Mr. Morgan's "reformatory movement" appears to me the most likely method by which it would begin its advance to a better system of marriage' than 'communal marriage.'

But what gave the impulse to the hypothetical moral reformation? Contact with a more advanced tribe is reckoned improbable by Mr. Fison (for how came the other tribe to be more advanced?), and so the moral impulse 'must have been derived from a higher power,' from the Good Spirit, or from ancestral spirits, as in the myths of the Dieri, the Woeworung, also of the Menomini Redmen of North America, a branch of the Algonquins; and the Euahlayi tribe.

According to the Menomini, there is, or was, a Being who 'made the earth.'[104] His name being interpreted means 'The Great Unknown,' but only extreme believers in the theory of religious borrowing will say that he was Sir Walter Scott, Bart. He (The Great Unknown) created 'manidos or spirits,' in the shape of animals, or birds. The chief birds (as often in Australia) were Eagles and Hawks. The Bear 'came out of the ground,' and was turned into an Indian, by the Great Unknown, alias 'The Good Mystery.' He and the Beaver headed totem kins now in 'The Big Thunder phratry.' Other animals came in; there are now Bear, Eagle, Crane, and Moose 'phratries,' each containing a number of totems. All the people of a totem name in the Menomini tribe are akin to persons of the same totem in other tribes, say of the Sioux.[105]

These myths favourably illustrate the piety of the Dieri,

---

[104] I know that many students will decline to admit that there is such a myth of a Maker.
[105] Report of Bureau of Ethnology, 1892-1893, pt. i. pp. 32-43.

Woeworung, Euahlayi men, and Menomini. Like Mr. Fison (at one time, and 'under all reserves') these tribes leaned to the hypothesis of divine or supernormal intervention in matters totemic. The Dieri may be right, but a less difficult hypothesis is that there was never 'an undivided commune,' in the sense of Mr. Morgan and Mr. Fison, and that, consequently, it never was 'divided into exogamous clans.' If so, no miracle is needed: Nec Deus intersit nisi dignus vindice nodus. My own scheme needs no divine aid, nor deliberate legislation, 'in the beginning.' But that such legislation has intervened later, I think probable, or certain.

Messrs. Spencer and Gillen write: 'Rigidly conservative as the native is, it is yet possible for changes to be introduced.... There are certain men who are respected for their ability, and, after watching large numbers of the tribe, at a time when they were assembled together for months to perform certain of their most sacred ceremonies, we have come to the conclusion that, at a time such as this, when the older and more powerful men from various groups are met together, and when day by day, and night by night around their camp fires, they discuss matters of tribal interest, it is quite possible for changes of custom to be introduced.'[106] The Arunta traditions allege that such changes introduced by men of weight, and accepted after discussion, have been not unusual.[107] This is highly probable, now, but not in the beginning.

The Arunta historical traditions are of little value as to historical facts,[108] but the consciousness of the Central Australian tribes accepts the possibility that new customs may now be proposed, debated, and adopted. If no such thing ever occurred, the belief in its possibility could scarcely have arisen among the Arunta. But the possibility has its limits, and one of these is the deliberate primeval introduction of exogamy, for no conceivable reason, and its imposition on a society already totemic but not yet exogamous. Perhaps few critics will frankly say that exogamy was thus imposed; they will try to qualify or evade so improbable and antiquated a theory. Yet they cannot but slip back into it, while they believe in 'segmentations' of 'an undivided commune,' and of later totemic 'subdivisions' of the 'segmentations.'

In any case these Arunta and cognate tribes of similar usages, so recently discovered, so anomalous, so odd, are 'the only begetters' of the latest hypotheses of Mr. Frazer and Mr. Spencer—

---

[106] Natives of Central Australia, pp. 12-15.
[107] Ibid. pp. 15, 421-422, also p. 272.
[108] Here I dissent from Mr. Frazer and Messrs. Spencer and Gillen; the point is discussed later.

namely, that totems, originally, were co-operative industrial groups with no influence on marriage rules. Do the Arunta, then, present a surviving model of primitive Totemism, in other regions modified and contaminated; or is their Totemism not, like their metaphysics and psychology, a 'freak,' an unique divergence from the normal development, as I have from the first maintained?[109] All these difficulties and confusions, as to 'phratries' and totems, inevitably arise from the doctrine that the original totem groups were not at first exogamous, and only became exogamous when separate sets of them were scheduled under the two more recent exogamous primary divisions, or were segmented out of them. In that case it is not easy to see how we can escape from the impossible theory that exogamy, and the primary divisions, were the result, of direct legislative enactment. Even if we could believe this, we see no conceivable motive, except Mr. Fison's divine intervention, an idea which, it appears, he put forward quite provisionally in an argument with Lord Avebury.[110]

THE ARUNTA

The case of these Central Australian tribes, in regard to Totemism and marriage prohibitions, is so peculiar that it demands particular notice. Mr. Frazer some years ago propounded the hypothesis that the Arunta tribe, especially, are the most 'primitive' of living peoples, are still in 'the chrysalis stage' of humanity, whence it would follow that their singular kind of Totemism, and of marriage rules, is nearest to the beginning, and best represents the original type.[111] The Arunta, dwelling in the arid regions of the centre, have certainly been little contaminated by European influences. They are naked, houseless, non-agricultural nomads, like all the Australian tribes, and it is asserted by Messrs. Spencer and Gillen and others that they have not yet discovered the rather obvious facts as to the reproduction of the species. All this has certainly a primitive air. But I have ventured to reply that the Arunta, as regards the family, are confessedly more advanced towards individual marriage than their neighbours, the Urabunna, with whom they freely intermarry.[112] Unlike what is told of the

---

[109] Fortnightly Review, June 1889.
[110] In 1895, J. A. I. xxiv., no. 4, p. 371, Mr. Fison abandons hope of a certain discovery of the origin of exogamy.
[111] Fortnightly Review, April, May, 1899.
[112] Spencer and Gillen, pp. 68, 69, 121.

Urabunna, the Arunta recognise 'individual marriage.' They deliberately and ingeniously modify their system on the occasion of intermarriage with the Urabunna. These reckon descent in the female, the Arunta in the male line.[113] The office of Alatunja, or head man of a local group, among the Arunta, is hereditary in the male line, descending to a brother of the late Alatunja, if he leaves no adult son.[114]

Moreover, the Arunta, and cognate tribes, occupy an area of 750 miles, and their meetings and discussions last for months. A people truly primitive cannot be conceived as capable of such immense local associations, and of such prolonged and pacific assemblies. Again, Messrs. Spencer and Gillen, rightly or wrongly, believe that 'communal marriage' is the earlier institution, and that it persists, 'slightly modified' among the Urabunna, but not among the Arunta. Thus, beyond all doubt, the Arunta are more developed, more advanced, than the Urabunna, and it is hardly safe to say that, where their organisation differs from that of the Urabunna, and other tribes in general, it differs because it is more 'primitive.' It must be less primitive, a special divergence from the type.

ARUNTA METAPHYSICS

Again, as proof that they are in no chrysalis stage, the Arunta possess a reasoned theory of things, so ingenious and complex, so peculiar, so extraordinary, so carefully atheistic, that one could scarcely believe it possible for naked savages, were it not so well attested. The theory is that of the original evolution of types of life into plants and animals, which, with the minimum of extra-natural aid, became human. The human beings possessed souls, which on the death, or disappearance into earth, of the original owners, were hereditary, being reborn into Arunta children. These souls each of a given totem (the plant or animal or other thing which first became human) haunt certain local centres. One place is the centre of Grub totem souls, another of Cat totem souls, and so forth. Each new child is of the totem of the haunted locality where the mother supposes that she conceived it; a totem soul of that locality has become incarnate in her, and from her is born. Thus the wife may be of one totem, the child of another; the husband may be of the wife's totem, of the child's, or of another. The totem is thus no bar to marriage, and is not inherited, all this being the result of the

---

[113] Ibid. p. 70.
[114] Ibid. p. 10.

peculiar philosophic system of the Arunta. Their totemism is thus a 'sport,' and not the original form of the institution.

We cannot reverse the case, the philosophy of hereditary totemic souls cannot be the result of the present mode of inheriting the totem from the group of souls that haunts each locality, it cannot be a myth invented to explain that custom. That custom requires the peculiar Arunta soul-belief as its basis, and cannot exist without the belief. If the child received its totem name from the place where it is born, we might say, 'Originally the child was called after the place of its birth.' (Arunta children still receive territorial personal names from the place of their birth.) 'Later, Totemism came in with totem local names, each place having a totem title. The local totem name of the place where a child was born was then given to each child. Still later, arose a myth that totem souls haunted each place, and that the child received its totem name because a local totem soul was incarnated in it, at the place where it was born.' We cannot maintain this theory—which makes the present Arunta belief a myth to explain the present Arunta custom—because that custom it does not explain. The child receives its totem name, not from the place where it is born, but from the place where the totem soul entered into its mother. Nor can we assume that totem names were originally given, not to human groups, but to districts of territory. Thus the present Arunta mode of obtaining the totem, in each case, is the direct result of the Arunta philosophic belief. That belief is peculiar, is elsewhere unheard of, is the property of a tribe distinctly more advanced in marriage rules, and local solidarity, than some of its neighbours, and therefore cannot be primary. It follows that the Arunta mode of obtaining the totem, not by inheritance, is not 'primitive,' is not the original model from which the rest of savage mankind has diverged. This I state, because, as a rule, a belief exists to explain an institution, and, as a rule, an institution is not the result of a belief.

ARUNTA TOTEM EATING AND TRADITIONS

Each Arunta totem kin may now eat, in moderation, of its own totem, and each kin does magic (Intichiuma) for the benefit of its totem, as part of the food supply of the tribe in general. The traditions represent men and women of the same totem as, of old, usually intermarrying (that is, as endogamous): while they are also said, as a rule, to have fed almost exclusively on their totems, being thus endophagous.

All these usages, real or traditional (except doing magic for the benefit of the totem), are at the opposite pole from the customary exogamous and exophagous Totemism of savage tribes all over the world, and even in Australia. If, therefore, the Arunta and tribes practising the same usages are primitive (it may be, and has been argued), their Totemism is, in origin, the earliest known case of the division of labour; each group selecting and working (by magic) for the benefit of its totem, as part of the tribal food supply. I elsewhere argue that each group must probably have had a recognised connection with its totem, before it set out to do magic for the propagation of the creature.[115] But I have also maintained that the Arunta are far from being 'primitive,' but are rather a 'sport,' and that their usages represent a local variation from the central stream of Totemism; not Totemism in its earliest known form.

DR. DURKHEIM ON THE ARUNTA

I had written on this topic in the Fortnightly Review (June, 1899), and in another chapter of this book ('The Origin of Totemism'), before I saw the essay of Professor Durkheim, of Bordeaux, Sur le Totémisme.[116] It is encouraging to find that Dr. Durkheim, independently, has worked out the same theory—namely, that the Arunta are not in the primitive stage of Totemism, but represent a very peculiar divergence from the type, and that their historical legends (more or less accepted by Mr. Frazer and Mr. Spencer) are mainly myths, told to account for certain facts in their social arrangements. We are not to reason from their single case, says Dr. Durkheim, as against the great mass of our knowledge of Totemism and totemistic exogamy and exophagy. 'In place of being a perfectly pure example of the totemistic régime, is not Arunta Totemism a later and disfigured (dénaturée) development?' For many reasons, says Dr. Durkheim, 'the Arunta are among the most advanced of the Australian peoples,'[117] and he gives his grounds for this opinion, some of which I had already adduced in 1899. Entering into detail, Dr. Durkheim readily shows that, though the Arunta now permit marriage between persons of the same totem (which is not hereditary on either side, but casual), they are, for all that, exogamous, in a fashion resulting from precise Totemism in their past.

---

[115] See 'The Origin of Totemism,' infra.
[116] L'Année Sociologique, 1900-1901, pp. 82-121.
[117] Ibid. v. 89-90.

They may not marry within the two primary divisions (which Dr. Durkheim styles 'phratries'). Each phratry contains two (sometimes four) other 'classes' (exogamous), and phratries arose in the combination of 'two elementary exogamous totem groups'—as I have already suggested. Now phratries, we have agreed with Mr. Howitt and Mr. Fison, were, in all probability, themselves originally totemic. Mr. Frazer also says, 'We should infer that the objects from which the Australian phratries take their names were originally totems. But there seems to be direct evidence that both the phratries and subphratries actually retain, in some tribes, their totems.'[118] If the opinion be correct, the phratries of the Arunta, which regulate their marriages, were originally local totem groups. On my system, then, namely, that totem kins were originally, or very early became, exogamous, were exogamous before 'phratries' arose, and before the so-called 'bisection' was made, then the Arunta organisation was originally that of exogamous Totemism. At first, though not now, totems regulated Arunta marriages.

Dr. Durkheim, in the passage cited, says that the two exogamous phratries are composed of 'two elementary totem groups, également exogames.'[119] Dr. Durkheim, who here is of my opinion, writes, 'It is not true that, among the Arunta, the totem has always been' (as it is now) 'without influence on marriages, nor, above all, is it true that Totemism, generally, implied endogamy.' Yet, according to Arunta myth, the ancestors of the 'dream-time' (Alcheringa) were endogamous, as a general rule, and, as a general rule, were endophagous, ate their totem animals or plants. The ancestors of their traditions fed on their own totems, 'as if by a functional necessity,' say Messrs. Spencer and Gillen. But this simply cannot be true, for each totem is not in season, (plums, for instance), or accessible, all the year through, and, if it were, it would be exterminated by endophagy. The traditions, again, do not represent the men of the totem groups as really and religiously endogamous. They exercised marital privileges, not only over the women of their totem group, but over any other woman they could come across. Certain totem groups are represented in the legends as wandering across the land, the men living with women of their totem group, while 'there is nothing to show definitely that marital relations were prohibited between individuals of different totems.' The men accepted the caresses of such women of other totems as they encountered; but their habitual mates were the women of their

---

[118] Totemism, p. 83.
[119] L'Année Sociologique, v. 92.

own totem.[120] In the alleged state of perpetual trek, the wives were naturally, in the opinion of the myth makers, of the group. At present an Arunta marries in or out of his totem; as he pleases.

THE RELATIONS OF TOTEMS AND 'PHRATRIES' AMONG THE ARUNTA

The relations of the totem groups to the 'primary divisions,' or 'phratries,' among the Arunta and cognate tribes, are, as we have already stated, entirely peculiar. We have seen that, in North America, and in Australia generally, no phratry ever contains the same totems as its linked phratry, and we have seen that Mr. Frazer calls this the natural arrangement.[121] If so, the present Arunta arrangement is not natural; it is a divergence from the natural type. Among the Arunta, 'no totem is confined to either moiety' ('phratry') 'of the tribe.' There is only 'in each local centre a great predominance of one moiety.'[122]

Dr. Durkheim regards the present state of Arunta affairs (the totems not being peculiar to either phratry) as une dérogation. Originally, he thinks, as among the Urabunna, each phratry contained only totems which were not in the other phratry; and he detects survivals, among the Arunta, of the earlier usage. At present the Arunta totems show 'a slight tendency to skip' (chevaucher) 'from one into the other phratry, doubtless because the Arunta totem system is no longer complete'—and no wonder, as Arunta totems are now not hereditary, but derived from the totem souls haunting each locality. Again, in Arunta legend, the ancestors 'were divided into companies, the members of which bore the same totem name, and belonged as a rule to the same moiety' ('phratry') 'of the tribe,' as now among the Urabunna, 'who are in a less developed state than the Arunta.' So say Messrs. Spencer and Gillen, and thus Arunta legend points to a past in which Arunta usage was, in this matter, as a rule the same as that of the less developed Urabunna: which I believe it really was.

But we can hardly accept the legends when they fit, and reject them when they do not fit, our theory! I lay no stress on the legends.

If, however, the Arunta 'phratries' originally, as Dr. Durkheim and I believe, never contained the same totems, then each Arunta totem group was, at that time, necessarily exogamous. No man or

---

[120] Spencer and Gillen, p. 419.
[121] J. A. I., N.S., i. 285.
[122] Spencer and Gillen, p. 120.

woman could then marry within the totem, as, at present, the Arunta can and do. They were barred by the phratry limit: persons of their totem were never in the phratry into which alone they could marry. So no one then could marry a member of his or her own totem kin. 'It is, therefore, untrue that marriage has always been permitted between members of a totem,' says Dr. Durkheim, though Arunta legend declares for the opposite view.

ARUNTA MYTHS

Here I am apt to agree with Dr. Durkheim. The evidence of the Arunta legends as to the customs of the Alcheringa, or 'dreamtime,' is 'such stuff as dreams are made of.' The legends are 'statements, invented mainly by popular fancy,' says Dr. Durkheim, 'to explain existing institutions, by attaching them to some mythical beings in the past. They are myths, in the proper sense of the word.' They are not marked by authenticity.

Against this idea we have the opinion of Mr. Frazer, and of Messrs. Spencer and Gillen.[123] The Arunta traditions, they say, and Mr. Frazer agrees with them, do not explain the present system, but deal with a former state of organisation and with customs quite different from the present. They do, but the Arunta invented the customs described in their myths, on purpose to explain, mythically, how the present customs arose out of deliberate modification of the alleged older customs. Messrs. Spencer and Gillen themselves assert this: 'the traditions point to a very definite introduction of an exogamic system, long after the totemic groups were fully developed, and, further, they point very clearly to the fact that the introduction was due to the deliberate action of certain ancestors,' which is the theory of Mr. Lewis Morgan!

The rest is true, but I, like Dr. Durkheim, conceive that all is (except where we have external evidence for deliberate modification of the 'classes') merely part of the Arunta ætiological or explanatory myth. That myth starts from the belief (Mr. Howitt's belief?) in primary totemic, but not exogamous groups, such as are precisely the present groups of the Arunta, though not of their neighbours the Urabunna, or of totemists in general. This exceptional condition of Arunta affairs needed explanation, and got it, in the myth that the groups were originally totemic, but not exogamous, as Arunta totem groups still are. Exogamy (not applying to totem groups, but to 'phratries') was brought in, the myth says, by deliberate action, by

---

[123] J. A. I., N.S., i., nos. 3, 4, p. 276.

our old friend, 'the Legislator,' The Arunta traditions, therefore, do explain 'the origin of the present system,' of the Arunta, as far as exogamy goes; and their explanation is as much a speculative hypothesis as Mr. Morgan's equivalent theory. It is one more example of the coincidence of savage myth and scientific hypothesis.

MR. SPENCER ON ARUNTA LEGENDS

I understand Messrs. Spencer and Gillen to contest this opinion, in one passage, and to assert it, under qualifications, in another. Their exact words must be given. 'If they' (Arunta traditions) 'simply explained the origin of the present system out of, as it were, no system, then we might regard them as simply myths invented to account for the former' (i.e. 'the present system'), 'but when we find that they deal with a gradual development, and with a former state of organisation and customs quite different from, and in important respects at variance with, the organisation and customs of the present day, we are probably right in regarding them as actually indicative of a time when these were different from those now in force.'[124]

Now to what do the traditions amount, as regards earlier marriage laws and customs at variance with those now in use among the Arunta? They amount to this: (1) Men of one totem had marital relations normally with women of the same totem. It is no longer the case that Arunta men have relations, normally and exclusively, with women of the same totem; a man may marry a woman of his own totem, or not, as he pleases. But so, in the traditions of the primeval trek, a man might, and did, take women of other totems as he pleased, by conquest probably; though these women seem to have lived, hitherto, solely with men of their own totem. The tradition starts from the hypothesis that all members of each mythical wandering totem group were originally of the same totem. That being so, the men naturally lived, when on trek, with women of their totem, taking women of other totems as they came across them. No longer on trek, the Arunta of to-day do the same thing, many women of their own or any other totem. The only shade of difference arises from the nature of the mythical theory, that many totem groups were originally migratory. But the present Arunta system of 'go as you please' in marriage (as far as totems are concerned) differs from the regular custom of the neighbouring

---

[124] J. A. I., N.S., i. 276-277.

Urabunna, for example. That difference, the Arunta probably feel, needs explanation. So their myth explains it, 'we Arunta always acted thus from the beginning.' So far the 'tradition' of Messrs Spencer and Gillen seems to me to be an ordinary explanatory myth.

(2) At the supposed time (a time when many human types were still in the husk!) men and women of what are now 'exogamic groups' ('phratries' or 'classes') had marital relations contrary to present usage.

But did the phratries or classes then, according to tradition, exist at all? The legend says that the men of the Little Hawk totem had these 'phratries' and classes, Kumura and Purula and so on (the names then carrying no known exogamous prohibition, as now, for the legend does not say that these 'classes' were exogamous). The Little Hawk men had arrived at the arts of making flint knives, and using them in circumcision. This they taught to less advanced groups, who tooled with fire sticks. But they only let their pupils have 'very rough' stone knives (Palæolithic, probably), at first. 'It was these Little Hawks,' say our authors, 'who first gave to the Arunta the four "class" names. We may presume that along with them there was instituted some system of marriage regulations, but what exactly this was there is no evidence to show.' Either the Little Hawks introduced exogamy, or they did not, a valuable result of traditional evidence.[125] 'As yet we have no indication of any restrictions with regard to marriage as far as either totems or classes are concerned,' say Messrs. Spencer and Gillen. Then why does the legend aver that the class names existed? Why did they exist? Now the existing restrictions of the 'classes' need explanation, and get it, from the myth; but, as there are no Arunta totem restrictions on marriage, at present the myth naturally says nothing about them. At this mythic period, 'persons of the Purula and Kumura classes, who may not now marry one another, are represented as living together.'[126] (3) Next 'the organisation now in vogue was adopted.' But, in its first shape, due to the wisdom of Emu men, it permitted marriages, which are now (4) forbidden by the superior intelligence of men dwelling further north, 'and it was decided to adopt the new system,' that is, the present Arunta 'class' system.

Now the Arunta are still accepting innovations from the North, and this part of the myth need not be mythical.

---

[125] Native Tribes of Australia, pp. 396-402, 421.
[126] Native Tribes of Australia, p. 418.

But the whole traditions, full of stark mythical inventions (including a myth like that of Isis and the mutilation of Osiris), amount merely to this. Society was totemic, but the totems were not exogamous; rather endogamous of the two. Society among the Arunta is still totemic, but not, as far as totems go, exogamous. In this it differs from the usual rule, and the myth explains why,—'it was always so.' But Arunta society is exogamous as regards the 'phratries' and classes, and that has to be explained by the myth. The myth therefore explains by saying that Emu men introduced a deficient, and northern men an adequate, system of exogamy—that which now prevails. Messrs. Spencer and Gillen, however, appear to deny that the 'traditions' 'simply explain the origin of the present system, out of, as it were, no system. It is true that the traditions do give stages in the arrangement of the present system; but they also do 'explain the origin of the present system.' And Messrs. Spencer and Gillen not only admit this, but, as we saw, even think the explanation 'quite possible.' The explanation, I repeat, is that the system 'is due to the deliberate action of certain ancestors,' Emu men and wiser Northern men.

Of course, as we tried to show, that explanation of primeval exogamy is improbable, but it is the explanation given by the Arunta legend. With a grain of fact, as to innovations from the North, the legend is a myth, an ætiological myth, a myth explanatory of the origin of the present organisation. History it is not. The Arunta 'traditions' are not historical evidence in favour of the new hypothesis that the Arunta are 'primitive,' are in 'the chrysalis stage' of humanity; (this they deny): that Totemism, in origin, was a magical co-operative and industrial association; that the original totems were not exogamous; and that exogamy was superimposed by legislation, or grew out of an organisation so imposed on a society of non-exogamous totem groups. Whatever the value of that hypothesis, it has no historical support from the Arunta traditions. History is a very different thing.

The Arunta still marry, at pleasure, in or out of the totem, merely because their totems are now scattered about among their exogamous divisions. This is not the 'natural arrangement' (as Mr. Frazer assures us), is not the inevitable original arrangement, and is not the case with their neighbours, the Urabunna, who are confessedly 'less developed than the Arunta.' The Urabunna system, therefore, is more archaic, ex hypothesi than that of the Arunta, which must be less archaic. It is, I repeat, peculiar, isolated, needs explanation, and the Arunta traditions give the explanation. The ancestors took women in or out of the totem, as at present the Arunta do; exogamy by classes was later imposed, says the myth.

Dr. Durkheim appears here to hold the more logical position. There was, I conceive, with Dr. Durkheim, and have stated, though Messrs Spencer and Gillen and others deny it, 'a primary relationship between the totemic system and exogamy.'[127]

CHAPTER IV

ARUNTA PHRATRIES AND TOTEMS

The essential question is, why, among the more archaic Urabunna, do the large exogamous divisions never include the same totems, whereas, among the more highly developed Arunta, they do? If we can show how the Arunta, if once organised on the Urabunna and North American model, came to slip out of it; while we cannot show how the Urabunna, and most other tribes, if once on the Arunta model, came to desert it (as they must have done), then it will seem probable that the Urabunna organisation, the regular universal Australian organisation, is the older.

The sequence of events, as understood by Messrs. Spencer and Gillen, was this, or, at least, may thus be conceived. We take two tribes, say Urabunna and Arunta. They both have many totem groups, totemic, because (on this theory) each group had, for its 'primary function,' the working of magic for the object which was its totem. The totem had primarily, on this theory, no relation to marriage rules. It is 'quite possible' that certain persons then deliberately introduced exogamous divisions.... 'so as to regulate marital relations.' The exact purpose, however, is unknown; 'it can only be said that far back in the early history of mankind, there was felt the need of some form of organisation, and that this gradually resulted in the development of exogamic groups.' This position I have already criticised; it is not intelligible to me. However—the exogamous division was made, and then all the totems might be arranged separately in the two divisions, by the Urabunna, 'and perhaps the majority of Australian tribes' (and the American tribes) or, 'this was not done,' as by the Arunta. Consequently, Messrs.

---

[127] Op. cit. p 279.

Spencer and Gillen think, the rule which prevents an Urabunna man from marrying a woman of his own totem, has nothing, primarily, to do with the totem, but is a mere inevitable consequence of the system which, among all tribes but the Arunta, excluded each totem from one of the two exogamous divisions, and placed it (not among the Arunta) in the other. My own system—I need not reiterate it—is the reverse of all this.

The Arunta, I contend, probably had, originally, the usual organisation, but have lost it for obvious reasons, so that now the same totem may occur in both of the large exogamous divisions, and persons of the same totem may now intermarry.

The traditions of the Arunta represent the exogamous 'phratries' as later than the totemic (but not yet exogamous) division. Dr. Durkheim thinks this improbable or impossible. It is true that the 'phratries' or 'classes' are now much more important, among the Arunta, than the totems, on which Dr. Durkheim insists. They need not, therefore, be earlier.

VIEWS OF DR. DURKHEIM

The theory of Dr. Durkheim is not, perhaps, expressed with his usual lucidity; at least I have found some difficulty in understanding it. The following summary, however, seems to be correct. 'The phratry,' he says, 'began by being a clan' (in my terminology an exogamous local totem group). 'There is no reason why this general idea should not apply to the Arunta. Consequently, since there are actually two exogamous phratries, we have reason to admit that this society was originally formed by two primary clans, or, if any one prefers the phrase, by two elementary totem groups, both exogamous (également exogames), for under this form the two phratries must have begun to exist. Now in that case there was at least a moment when marriage was forbidden between members of the same totem,' though now among the Arunta this rule no longer obtains.[128]

So far Dr. Durkheim and I hold identical views; we differ on a point of detail. What are, and whence came, the totems within the phratries? Dr. Durkheim conceives the case thus: Originally there was a 'clan' (local totem group) which was exogamous, and married out into one other equally exogamous clan. The members of each such exogamous totem group ('clan') then multiplied and 'swarmed off,' in colonies, and all such colonies took a new totem, while

---

[128] L'Année Sociologique, v. 91, 92.

retaining 'the sentiment of their primary solidarity' with the original totem group. These are the 'secondary' totem kins. But why should they take new totem names and new totems?[129] I know not, but the original group from which they swarmed off now became their 'phratry.' This phratry, in many cases, still has a totem name, 'which is the proof that it is, or has been a clan,' that is an exogamous totem group.[130] Therefore exogamous totem groups were 'primary,' the existing totem kins are 'secondary,' they have split off from the original groups. As far as I am able to follow Dr. Durkheim's reasoning, he and I differ on this one point. We both regard the two 'phratries' as having been originally local exogamous totem groups, which united in connubium. But in each 'phratria' there exist several totem kinships. Dr. Durkheim regards these as 'secondary' branches which split off from the two original local totem groups, and which, in each case, took new totem names, while retaining membership in their original totem groups, now 'phratries.' They are totemic colonies of a totemic metropolis. I, on the other hand, as has been explained, conceive that each of the two local totem groups which became phratries (say Emu and Kangaroo) already, by the action of exogamy in a region where there were many totem groups, and by virtue of female descent, contained within it persons who were of various totem kindreds. Dr. Durkheim, on the contrary, seems to think of the existence of but two primal exogamous clans in a given region. Groups emigrating from these took new totem names, while retaining the phratry name and connection with their mother clans, now phratries.

Why the clans were totemic at all does not appear. I understand that they were exogamous out of respect for the blood of their totems, the totem tabu (p. 57, note I).

Against the hypothesis it may be urged (1) that we do not know that emigrants from a local centre ever select new totem names—unless, indeed, they reach a region where their old totem does not exist. This cannot have occurred constantly. Again (2), Dr. Durkheim's theory involves the same difficulty as my own. How did the colonies from the Kangaroo group happen never to select the same totem as colonies from the Emu group, so that the same totem never occurs in both phratries? This implies deliberate arrangement. If however, totem names were given from without, by neighbours (as I shall argue), the case could not occur at all, and the same totem would appear in both phratries.

---

[129] This idea we shall find again later, in another part of Dr. Durkheim's system.

[130] L'Année Sociologique, i. 6, 7.

If we adopt the hypothesis that two friendly 'families,' or 'fire circles,' of a cousinly character, set the first example of exogamous intermarriage—exclusively with each other—and then got totem names, they might become phratries, but whence arose the totem kins within the phratries? Shall we say that other such 'families,' increasing in size, and receiving totem names, came in, two by two, to Emu and Kangaroo, each of the new linked adherents taking opposite sides, Opossum going to the Kangaroo, Bandicoot to the Emu phratry? This would give the totems within the phratries, by a constant accession of other pairs of phratries, which subordinated themselves, one to Emu, one to Kangaroo. Either this hypothesis, or Dr. Durkheim's, or my own, accounts for the phratry plus totem kins arrangement, without supposing the deliberate bisection of a hitherto undivided commune. That hypothesis, if any one, of the other three, Dr. Durkheim's, my own, or the theory of accessions to the pair of exogamous intermarrying families, be accepted, is therefore not forced upon us in defect of a better.

HOW DID THE ARUNTA ANOMALY ARISE?

At all events, the Arunta 'clan' (totem kin) is now no longer exogamous, and two Arunta phratries can now contain members of the same totems, contrary to Kamilaroi, Dieri and Urabunna and American custom. How did this anomaly arise? Dr. Durkheim supposes that the change began when Arunta kinship came to desert the female and to be reckoned in the male line. This appears to Dr. Durkheim to be indicated by the complicated and ingenious arrangements made when an Urabunna (who reckons by the female line) intermarries with an Arunta, who reckons by the male line.[131] These arrangements, he thinks, are no novelty devised for the occasion: the Arunta merely revert to their old way of reckoning by the spindle side. When the Arunta changed their system, and reckoned in the male, not, as of old, in the female line, the children now belonged to the 'phratries,' not of their mothers, as previously, but of their fathers. Each 'phratry' then bartered a sub-class of its own for a sub-class of its partner. Each bartered sub-class thus brought its totems into the other 'phratry,' and there was no longer a totem group entirely peculiar to one or other 'phratry.' Consequently, a member of the Kangaroo totem could marry a woman of the same, if she were in the opposite 'phratry' to his own.

Might not the same results follow from the mere fact, that,

---

[131] L'Ann. Soc. v. 104-107; Spencer and Gillen, pp. 68-69.

among the Arunta, the totem is now inherited neither from father nor mother, but is derived simply from the totem souls that haunt the particular glen or hill where the child was conceived? By this means a totem soul can get into a child of the 'phratry' to which that totem did not originally belong, and thus the totems 'skip' from one 'phratry' to another, contrary to general rule in Australia and North America. This is the explanation of the Arunta anomaly which Messrs. Spencer and Gillen accept. 'The spirit child' (of the Lizard totem) 'deliberately, the natives say, chose to go into a Kumura' (class) 'woman, instead of a Bulthara woman.... Though the class was changed, the totem could not possibly be.... Owing to the system according to which totem names are acquired, it is always possible for a man to be, say, a Purula' (class) 'or a Kumura' (class) 'and yet a Witchetty; or, on the other hand, a Bulthara' (class) 'or a Panunga' (class) 'and yet an Emu' (totem). But, if he is thus born to a totem which was not originally (on my theory) a totem of his phratry, a man loses the chance of being an Alatunja, or head man of a local group.[132] Thus the Arunta anomaly arises merely and necessarily from the Arunta philosophy of souls. That philosophy is an isolated freak, and it has upset and revolutionised Arunta Totemism, which, therefore, is the reverse of the 'primitive' model.

CHAPTER V

OTHER BARS TO MARRIAGES

The prohibitions on marriage, with which we have hitherto been concerned, are based on what savages regard—while we do not—as relations of kindred. Men and women of the same 'phratry' or 'primary division' may not intermarry (where such divisions exist), nor may men and women of the same totem name. Civilised society, at least in Europe, now recognises no such things as the 'phratry' or the totem kin. When Mr. George Osborne, in Vanity Fair, was asked whether he was akin to the ducal House of Leeds, he

---

[132] Spencer and Gillen, pp. 125, 126. The reader is recommended to study Dr. Durkheim's passage cited in the last note, the topic being difficult.

replied that he bore the same arms—these having been conferred on his father by a coach-builder. In savage society, Captain Osborne's answer would have been satisfactory. He would really have reckoned as a kinsman of all other Emus, if his totem and badge (coat of arms) was an Emu. In Scotland the Campbell name used to be regarded as implying at least a chance that the bearer was of the blood of the Black Knight of Loch Awe, and had a right to the Campbell tartan, and badge, the gale, or bog-myrtle. But, of course, as a rule, in modern society, a common surname is no proof of kinship, and coats of arms are usually home by the middle classes, and peers of recent creation, without much inquiry.

So far, then, the totemic rules which prohibit certain marriages, have no resemblance to our own definite 'forbidden degrees,' based on nearness of blood. The savage rules, as they stand, include our notions of kindred, but these notions, as far as they are recognised, are not conterminous with ours. But the 'phratry' prohibitions, and the totem prohibitions, are not the only bars to marriage among such peoples as the Australians.

The other bars are lucidly described by Messrs. Spencer and Gillen.[133] 'There are still further restrictions to marriage ... and it is here that we are brought into contact with the terms of relationship.' We find that a woman may belong to a totem kin (and phratry) into which a man may lawfully marry, 'yet there is a further restriction preventing marriage in this particular case.' Thus a male Dingo (among the Urabunna) may marry a female Water Hen, as far as 'phratry' and totem are concerned. But he may not marry a woman of the Water Hen totem if she reckons (1) as his father's sister (i.e. of his father's generation), (2) if she is his child, or his brother's child (of the next generation), (3) if she be one of his mother's younger brother's daughters: but he may marry her if she (4) be one of his mother's elder brother's daughters. All women of that category (4) are Nupa, or nubile, as far as this man goes. In category I, the women (including 'paternal aunts,' as we reckon) are of an older generation than the man; in category 2 they are of a younger generation (including our 'children' and 'nieces'); in category 3 the women include our cousins on the maternal side, by uncles younger than our mothers, and, in category 4, they include our cousins on the maternal side, by uncles older than our mothers. We Europeans, being males, may not marry into categories 1 and 2, but if not Catholics, we may marry into categories 3 and 4; if Catholics, we may—if we can get a dispensation.

---

[133] Op. cit. p. 61.

In the Australian system the oddest thing is that a male may marry into what, in our phrase, includes his younger maternal uncle's daughter, but not his elder maternal uncle's daughter. But we here use the words 'uncle,' 'aunt,' and 'cousin,' only by way of illustration. The Urabunna, and tribes of their level generally, have no such words. All children (category I) 'of men who are at the same level in the generation, and belong to the same class and totem, are regarded as the common children of these men,' or, perhaps we should rather say, are called by the same name, Biaka, as a man's own children are styled. A man knows very well which children he reckons his own, though, as will be seen, he has little ground for his confidence. In the same way a child, though he calls all men of his father's class, totem, and level in the generation, Nia (fathers), knows well enough which Nia feeds him, pets him, thrashes him for his good, and, generally, plays the paternal part. For example, a man informs you that this or that native, by personal name Oriaka, is his Okilia, 'and you cannot possibly tell without further inquiry whether he is the speaker's own or tribal brother, that is the son of his own father, or of some man belonging to the same particular group' (by 'phratry,' totem, and seniority) 'as his father.'[134] But you can learn 'by further inquiry:' the actual relationship, in our sense of the word, is recognised.

'GROUP MARRIAGE'

These facts necessarily lead to the question, are all men of one class, totem, and seniority, actual husbands of all women of the opposite class, different totem, and equivalent seniority? (Group Marriage). Or, if this is no longer the case, was it once the case? and are these sweeping uses of names which include our 'father,' 'mother,' 'brother,' 'child,' survivals of such a stage, called 'Group Marriage'? This question is still undecided; good authorities take opposite views of the question, which has bred, in the past, much angry controversy.

MR. MORGAN AND THE CLASS SYSTEM

The arrangement by 'classes,' 'the classificatory system,' was first brought into scientific prominence by the late Mr. Lewis Morgan, an American gentleman affiliated to the Iroquois tribe, in

---

134 Spencer and Gillen, p. 57.

his very original studies of the names for degrees of kinship.[135] A great deal may be said, and has been said, especially by Mr. McLennan and Dr. Westermarck, against Mr. Morgan's ideas and methods, but his large and careful collection of facts is of high importance. On what he called 'the Malayan system,' one name denoting kin includes all my brothers, sisters, and cousins. Another name includes my father, mother, my uncles, aunts, and all the cousins of my father, mother, aunts, and uncles. The generation of my grandparents and their relations is included in a third name; a fourth covers my children and their cousins, and the grandchildren of my brothers and sisters, with their children, bear the same name, for me, as my own grandchildren. From the names Mr. Morgan inferred the existence of certain facts in the evolution of systems of kindred. Everybody of the same generation lived together, once, on his theory, in 'communal marriage,' brothers, sisters, and cousins. There was promiscuity between all men and women in the same generation. Of course this involves the converse of Mr. Atkinson's Primal Law, as Mr. Atkinson observes in his eighth chapter. In place of the prohibition of brother and sister union being the earliest of prohibitions (as in Mr. Atkinson's system), the rule that they must unite, caused, in Mr. Morgan's opinion, the earliest form of the human family.

DIFFICULTIES OF MR. MORGAN'S THEORY

Mr. Morgan's theory, it must be observed, landed him at once in the fallacy of supposing that prohibitions of marriage of kinsfolk were originally the result of 'a reformatory movement.'[136] We have seen that, granting, for the sake of argument, Mr. Morgan's premise of an original 'undivided commune,' Mr. Fison is also deposited in the same difficulty, and was once even inclined to regard a theory of intervention 'by a higher power' (the Dieri myth) as not necessarily out of the question, if marriage was once communal. To reform such marriage relations, he says, 'would be a step in advance so difficult for men in that utter depth of savagery to take, that they would not be able to take it, unless they had help from without. This might be given by contact with a more advanced tribe; but if all the tribes started from the same level, that impulse would be impossible in the

---

[135] Systems of Consanguinity and Affinity of the Human Family (1871); and Ancient Society (1877); earlier in The League of the Iroquois (1854).
[136] So Mr. Fison candidly states, and Mr. Morgan saw his work, and wrote an introductory essay. Kamilaroi and Kurnai, p. 99.

first instance, and must have been derived from a higher power.'[137] Mr. Fison, as we saw, has since expressed the opinion that the origin of exogamy is probably indiscoverable, but I cite again his early remark to prove his sense of the insuperable difficulty of Mr. Morgan's theory.

How were men in his hypothetical condition to know that there was anything to reform? It needed a divine revelation!

Mr. Morgan was himself aware of this difficulty, and tried to get out of it, by using Darwinian phrases about 'natural selection'—'blessed words,' but here unavailing. He was in the posture of Mr. Spencer, between direct legislation to introduce exogamy, and gradual evolution of exogamy, as the slow result of the felt need of 'some organisation,'—its nature and purpose unknown. Thus Mr. Morgan, speaking of communal marriage, and its results, says that 'emancipation from them was slowly accomplished through movements which resulted in unconscious reformation.' These movements were, first, the 'class' system, then the 'gens' (totem system), 'worked out unconsciously through natural selection.'[138] This means, if it means anything, that, by a freak or sport, some people did not marry in and in, that they unconsciously evolved the totem system, that they therefore throve, while others who married in and in, and did not evolve the totem system, perished, and so we have the results of 'natural selection.' But why did some people avoid the habit of marriages of near kin which was so general? The position is that of Dr. Westermarck, who adds an 'instinct,' developed by natural selection,[139] an idea which involves arguing in a circle.

Again, that peoples marrying in the communal way would die out has to be proved: science has no certainty in the matter.

In any case, Mr. Morgan presently deserts his opinion about slow unconscious reformation, and his natural selection. 'The organisation into classes seems to have been directed to the single object of breaking up the intermarriage of brothers and sisters, which affords a probable explanation of the origin of the system. But since it does not look beyond this particular abomination it retained a conjugal system nearly as objectionable....'[140] The reader sees that Mr. Morgan cannot keep on the high Darwinian level. He relapses on a supposed moral reform with a single object of things 'abominable'—to us—and 'objectionable'—to us. But how did the

---

[137] Kamilaroi and Kurnai, pp. 160-161.
[138] Ancient Society, pp. 49-50.
[139] Cf. The Mystic Rose, pp. 444-445. Westermarck, p. 352.
[140] Ancient Society, p. 59.

pristine savages find out that such things were 'abominable'? Presently the totem prohibition ('the gens') 'originates probably in the ingenuity of a small band of savages,' for the purpose of modifying marriage law, and the daring novelty 'must soon have proved its utility in the production of superior men.'[141] Here we have the legislation due to human 'ingenuity,' and natural selection comes in to aid and diffuse the system. Later 'the evils of the first form of marriage came to be perceived' (what were they?) and this led 'if not to its direct abolition, to a preference for wives beyond this degree. Among the Australians it was abolished by the organisation into classes, and more widely among the Turanian tribes by the organisation into gentes.' The Australians have 'gentes' (totem groups) quite as much as the 'Turanians' or 'Ganowanians,' and we have tried to show that totems are prior to 'classes.'[142] But the Australians 'abolished' a form of marriage by an 'organisation,' which implies deliberate legislation. From this difficulty of legislation, so early and so moral, no advocate of the 'bisection' of an undivided commune and of its 'subdivision' into totem 'phratries' and kins, can escape, however he may make a push at 'natural selection,' and gradual evolution.

MR. MORGAN ON TERMS OF RELATIONSHIP

These perplexities do not predispose us in favour of Mr. Morgan's theory of the terms of 'Relationship,' which we have illustrated by the case of the Urabunna. He himself takes the Hawaiian terms, which are to the same effect. In brief, all the men and women of a generation are' brothers and sisters,' all those of the prior generation are 'fathers and mothers,' all those of the following generation are 'children.' Now, if ever all the men and women of a generation married 'all through other,' promiscuously, these terms of 'relationship' would be in place. First, we are told, brothers and sisters in a family intermarried, and the process 'gradually enfolded the collateral brothers and sisters, as the range of the conjugal system widened.' And then 'the evils came to be perceived,' what evils, how perceived, we do not know, and Reformation set in. It definitely began with the Australian 'Bisection,' 'the organisation into classes' (really into 'phratries'), and about the difficulties of that theory enough has been said.

The reader will naturally ask, What is the original meaning of

---

[141] Ibid. p. 74.
[142] By 'classes' Mr. Morgan here seems to mean phratries.

the words now used by Hawaiians, and Urabunna, and others, for the relations in which our 'father,' 'son,' 'wife,' 'husband,' 'mother.' 'daughter,' 'brother,' 'sister,' are included? Do the words embracing our terms 'brother' and 'sister' in Hawaii, or elsewhere, imply procreation, and issue (as in Greek), 'from the same womb'? Among the Arunta they cannot mean procreation, if they do not even know (as Messrs. Spencer and Gillen tell us), that there is any such thing as procreation. 'A spirit child enters a woman,' that is all. In the times of this primeval ignorance, words for relationships could not imply bearing and begetting; they must have meant something else. Say that they meant relationships in point of seniority: 'my male elder,' 'my female elder,' 'my male junior,' 'my female junior,' 'my male coeval or friend,' 'my female coeval or friend,' 'the man I may marry,' 'the woman I may marry,' 'the woman or man I may not marry.'

If low savage names for relationships meant that (no doubt they do not, or not often) then they would undeniably prove nothing as to a system of communal marriage. A baby points to any man or woman and says 'pa' or 'ma,' without any theory of communal marriage. Thus philologists must first interpret for us the original significance of these savage names of relationships. Once given, they would last, whatever they originally implied. Dr. Westermarck has urged this point.[143] In the terms themselves there is, generally, nothing which indicates that they imply an idea of consanguinity.' 'Pa, papa' (father), ma, mama (mother), and scores of others, 'are formed from the earliest sounds a child can produce,' and 'have no intrinsic meaning whatever.' Dr. Westermarck gives a long list of such words, applied to 'fathers, and all the tribe brothers of fathers,' and the same for mothers, concluding 'that we must not, from these designations, infer anything as to early marriage customs.' He does not deny that other terms of relationship have roots of independent meaning, 'but the number of those that imply an idea of consanguinity does not seem to be very great.' In Lifu (Melanesia), the word for 'father' means 'root;' for 'mother,' 'foundation' or 'vessel;' for 'sister,' 'not to be touched;' for 'elder and younger brother,' 'ruler' and 'ruled.'[144] The terms for father and mother denote consanguinity; the others, customary law, and status.

If we only knew the meanings, say, of the Urabunna words for relationships, we should learn much. But the truly amusing fact is that Mr. Fison, for example, did not know the language of the natives, and thought that probably not six white men in Australia

[143] Westermarck, pp. 85, 96.
[144] Lord Avebury, Origin of Civilisation, pp. 442-449, 1902.

had an adequate knowledge, and an adequate access to the notions, of the tribesmen. Of these one had been initiated, and, like a gentleman, declined to break the oath of secresy.[145] This was in 1880. Things may have improved. But unless our authorities know the languages, where are we? We do know that seniority is indicated: Father's elder brothers are Gampatcha Kuka (Warramunga tribe).

Mr. McLennan thought that all these terms were 'terms of address,' used to avoid the employment of personal names, and Dr. Westermarck holds that 'there can scarcely be any doubt that the terms for relationship are, in their origin, terms of address.' Messrs. Spencer and Gillen, after impartial consideration, cannot accept this view, for Australia; where the terms are very numerous, and stand for relations very complicated, connected with the intermarrying groups, and with social duties. In addressing a person, his or her individual name (our Christian name) is freely used.[146] They believe that the terms can only be explained 'on the theory of the former existence of group marriage, and further, that this has of necessity given rise to the terms of relationship used by the Australian natives.' These opinions are shared by Messrs. Fison and Howitt. The former says, 'It must, I think, be allowed that the classificatory terms point to group marriage,' and though Bastian denies this, Mr. Fison supports his theory by the Dieri custom of allotting paramours (pinauru) to men and women, out of the sets which may intermarry.[147]

To this problem we return; meanwhile it may seem impertinent in mere ethnologists of the study to hint a doubt as to the conclusions of observers on the spot. Mr. Crawley, however, has no hesitations. The use of the terms of relationship, he thinks, does not testify to a past of 'Group Marriage,' or to a remoter past of promiscuity, but is 'the regular result of the primitive theory of relationship; the system codifies a combination of relation and relationship, "address," and age.' The terms in use 'do not in themselves necessarily point to a previous promiscuity, or even to a present group marriage,' as Messrs. Spencer and Gillen believe.

The point is one on which I almost hesitate to venture a decided opinion. Much seems to depend on the original sense of the various terms, and on that point, in the case of Urabunna, and many other tribes, we have no light. But often the terms do not express consanguinity at all. There seems to be no word for 'daughter' as

---

[145] Kamilaroi and Kurnai, p. 60.
[146] Spencer and Gillen, pp. 56, 57, 59.
[147] J. A. I., May 1895, p. 368.

distinct from 'son,' 'nephew,' and 'niece.' The grandfather maternal is Thunthie, and Thunthunnie is Urabunna for totem, so that it is tempting to guess that Thunthie means 'a sire of the maternal totem.'[148] Kadnini, again (I speaking), means grandfather paternal, grandmother maternal, and grandchildren.[149] These relationships imply duties and services. 'One individual has to do certain things for another ... and any breach of these customs is severely punished.' An Arunta of the Panunga class calls all Kumura men 'fathers-in-law.' He gashes his flesh if any one of his 'fathers-in-law' dies, and he drops his dead game if he meets any one of them. They all have that advantage over him.[150] Thus these terms of relationship—communal in appearance—really involve certain duties, rather than relations of blood and affinity. But emphatically the terms are more than mere terms of address, as in Mr. McLennan's theory.

But these are usages of the system as it stands to-day. Is there behind it an 'undivided commune,' as Mr. Morgan held; is there actual 'group marriage'? I am not apt to believe that there is. Language shows, in the terms of relationship, a group of 'Mothers' for each child; but, as Mr. Darwin remarks, 'it seems almost incredible that the relationship of the child to its mother should ever have been completely ignored, especially as the women in most savage tribes nurse their infants for a long time.' A man's mother is one, and must be known, though he calls many women by the same name as he gives to his mother. She is lumped, in the terms of relationship, in one term with all the women whom the father might legally have married, but did not. The son, in addressing or speaking of his mother, overlooks the 'one love which needs no winning,' and his term has reference only to the present marriage law of his tribe. That law 'codifies' the terms, they result from that law, and that law, again, is based, if I am right, on totem prohibitions, on the desire to keep marriage between people of the same generation, and on the rights and duties of the generations. These prohibitions, of phratry, 'class,' totem, and age, leave only a certain set of women marriageable to a certain set of men. The name of this set of women is Nupa to their coevals, Luka to the succeeding generation. There is no name for 'wife,' no name for 'mother;' there are only names expressive of customary legal status, itself the result of the existing rules. Whatever their original sense, they all now connote seniority and customary legal status, with its

---

[148] Spencer and Gillen, p. 60.
[149] Ibid. p. 66.
[150] Ibid. p. 75.

reciprocal duties, rights and avoidances. 'It is the system, and not group marriage, which has given rise to these terms of relationship,' says Mr. Crawley.[151]

But what gave rise to the system? Mr. Fison has told us. 1. 'The division of a tribe (community) into two exogamous intermarrying classes....' 2. 'The subdivision of these two classes into four,' or, he suggests, the amalgamation of two tribes. 3. 'Their subdivision into gentes distinguished by totems.'[152]

But all of this theory we have already declined to accept for reasons given, and mainly because it involves (as I try to show) deliberate primeval reformatory legislation—without any conceivable motive. Again, we cannot accept Mr. Fison's system because it involves the hypothesis that a tribe, or 'community,' large enough to feel the necessity of bisecting itself for social and moral purposes, existed at a period when the difficulties of commissariat, of food supply, and of hostility, could seldom, if ever, permit its existence. A tribe is, I repeat, a local aggregate of small groups become friendly: it is not a primeval horde which keeps on subdividing itself, legislatively, for reformatory purposes. What social cement kept such a primeval horde, such an 'undivided commune,' together; and how did the animal jealousy of men so near to the brutal stage fail to rend it into pieces? How was it fed? How can we imagine a human herd—how supplied with food, who knows?—wherein each male sees each other male approach what female he pleases, perhaps his own preferred girl, without internecine jealousy? I cannot imagine this indifference to love in such a primitive Agapemone; I cannot understand its economics; any more than I can guess why such a state of affairs ever seemed— to its members—'abominable' and 'objectionable,' and a thing to be reformed; yet they 'bisected' it, and 'subdivided' the segments, all in the interests of morality—such is the theory.

As for the good-humoured laxity which enables all men and women to live together matrimonially at random, Mr. Morgan found an example, as he thought, in the Punalua of the Hawaiians. The word Punalua, when observed (1860) by Judge Andrews, meant 'dear friend,' or 'intimate companion.' A man called his sister's husband (our 'brother-in-law) his 'dear friend,' and a woman styled the wife of her husband's brother (her sister-in-law), her 'dear friend,' or Punalua. This shows that relations-in-law were not 'Foes-in-law,' or, at least, that this was not the official view of the case. It really does not follow that all the wives 'shared their remaining

---

[151] The Mystic Rose, p. 476.
[152] Kamilaroi and Kurnai, p. 27, cf. p. 70.

husbands in common.' Judge Andrews thought that this happy family 'were inclined to possess each other in common.' That was only the Judge's theory, also the theory of the Rev. Artemus Bishop. Probably there was a great deal of genial license and indifference among loose luxurious barbaric people, living in 'summer isles of Eden,' where food and necessaries were ready made by benignant Nature.[153]

> Each shepherd clasped, with unconcealed delight,
> His yielding fair, within the Captain's sight;
> Each yielding fair, as chance or fancy led,
> Preferred new lovers to her sylvan bed.[154]

This is vastly well, and the poet adds, in a liberal spirit,

What Otaheite is, let England be!

It is very well, but it by no means represents, probably, the manners of primitive man.

'We may conclude,' says Mr. Darwin, 'from what we know of the jealousy of all male quadrupeds,... that promiscuous intercourse, in a state of nature, is extremely improbable.... The most probable view is that primeval man aboriginally lived in small communities, each with as many wives as he could support and obtain, whom he would have jealously guarded against all other men. Or he may have lived with several wives by himself, like the Gorilla, for all the natives agree that but one adult male is seen in a band; when the young male grows up a struggle takes place for mastery, and the strongest, by killing and driving out the others, establishes himself as the head of the community. The younger males, being thus expelled and wandering about, would, when at last successful in finding a partner, prevent too close interbreeding within the limits of the same family,' just as the other male did.[155]

This second view of Mr. Darwin's is much like the theory of Mr. Atkinson, and is very unlike Mr. Morgan's theory of a human horde, living in communal marriage, or group marriage. Mr. Darwin's idea, moreover, the primitive groups being small, does not encounter the economic difficulties raised by the hypothesis of the 'undivided commune.' The strongest male practically enforced exogamy, as far as he was able, and maybe conceived to have entertained no scruples as to connection with his daughters. Mr. Darwin admitted that 'the indirect evidence' for communal

---

[153] Ancient Society, pp. 427-428.
[154] Captain Cook, of His Majesty's ship The Endeavour.
[155] Descent of Man, ii. 362, 363. Dr. Savage, Boston Jour. of Nat. Hist. v. 423.

marriage, and fraternal incest, was 'extremely strong,' but then 'it rests chiefly on the terms of relationship which are employed between members of the same tribe, implying a connection with the tribe alone, and not with either parent.' If, however, we have successfully explained these terms of relationship as not usually meaning degrees of consanguinity, but of customary legal status, under the prevalent customary law, the evidence which these terms yield for promiscuity, or group marriage, is extremely weak, or is nil, above all if our theory of how the legal status arose is accepted. And, if it is not accepted, back we come to primeval 'reformatory movements.'

In Lifu, the word for 'sister' means 'not to be touched,' land this is a mere expression of customary law. A man 'must not touch' any one of the women of his generation whom the totem tabu and the rule of the exogamous 'phratry' (in origin, we suggest, totemic) forbid him to touch. All such women, in a particular grade, are his sisters. Many women, besides his actual sisters, stand to him in the degree thus prohibited. All bear the same name of status as a man's actual sisters bear, but the name does not mean 'sisters' at all, in our sense of that word: namely, daughters of the man's real father and mother. It means tabued women of a generation. If the 'classificatory' terms which include our 'fathers,' 'sisters,' 'wives,' and the rest meant what our 'fathers,' 'sisters,' 'wives,' and so on mean, then the evidence from the terms, for communal or group marriage, would really be 'extremely strong.' But, as Messrs. Spencer and Gillen say, 'unless all ideas of terms of relationship as counted among ourselves be abandoned, it is useless to try and (sic) understand the native terms.'[156] Yet the whole force of the argument for communal marriage derived from savage terms of relationship rests precisely on our not 'abandoning' (as we are warned to abandon) 'all ideas of terms of relationship as counted among ourselves.'

The friends of group and communal marriage, it seems to me, keep forgetting that our ideas of sister, brother, father, mother, and so on, have nothing to do (as they tell us at certain points of their argument) with the native terms which include, indeed, but do not denote these relationships, as understood by us. An Urabunna calls a crowd of men of his father's status by the same term as he calls his father. This need not point to an age when, by reason of promiscuity, no man knew his father. Were this so, a man of the generation prior to his father might be the actual parent of the speaker, and all men under eighty ought to be called 'father' by

---

[156] Spencer and Gillen, p. 65.

him—which they are not. The facts may merely mean that the Urabunna styles his father by the name denoting a status which his father shares with many other men; a status in seniority, 'phratry,' and totem. We really cannot first argue that our ideas have no relation to the terms employed by savages, and then, when we want to prove a past of communal marriage, turn round and reason as if our terms and the savage terms were practically identical. We cannot say 'our word "son" must not be thought of when we try to understand the native term of relationship which includes sons in our sense,' and next aver that 'sons in our sense, are regarded as real sons of the group, not of the individual—because of a past stage of promiscuity making paternity indiscoverable.'

As Messrs. Spencer and Gillen say, we must 'lay aside all preconceived ideas of relationship,' when we study the Urabunna or other classificatory terms of relationships.[157] Let us do so, and the evidence borne by these terms to a past of communal marriage vanishes at once. That the terms often denote status in customary law is demonstrated. 'There are certain customs which are enforced by long usage and according to which men and women of particular degrees of relationship may alone have marital relations, or may not speak to one another, or according to which one individual has to do certain things for another, such as providing the latter with food, or with hair, as the case may be, and any breach of these customs is severely punished. The elder men of each group very carefully keep alive these customs, many of which are of considerable value to themselves....'[158]

Thus, you have speared a fish, or an opossum, but if you meet any man of your father-in-law's set, you must drop your spoil and make off. Consequently, I venture to take it, the terms of relationship in no way answer to our ideas of kin, but merely denote legal status.

HOW THE TERMS OF RELATIONSHIP ORIGINALLY AROSE

We cannot, as a rule, recover (or Australian students have not recovered) the original sense and etymology of terms like Biaka, Nia, Nupa, and so forth. We are thus left to choose between two competing theories of their nature and diffusion. If we advocate the hypothesis of consanguine marriage and group marriage, we must

---

[157] Op. cit. p. 67.
[158] Spencer and Gillen, pp. 67, 68.

suppose that the members of the 'undivided commune' of the theory, had once names absolutely identical in sense with our 'father,' 'mother,' 'sister,' 'brother,' 'son,' 'daughter,' and so forth. But the speakers, in each case, were obliged to apply these words with the utmost laxity, because who knew who A's father might be, and whether C's sister were really his sister or not, while every girl was the wife of every male of her generation, not barred by other laws, and so on? The promiscuity of living, then, made this lax use of words for relationships inevitable.

This is the usual hypothesis, and the sweeping scope of savage words for human relationships is accepted as proof that consanguine and group marriage once existed and left their marks in language. On the other hand, if communal marriage prevailed, the people who lived in that condition could not possibly have had ideas equivalent to our father, son, daughter, brother, wife, and so on. Our ideas of these relationships could not enter the human mind, at the hypothetical stage of culture when nobody knew 'who is who' and the hypothesis is wrecked on that fact.

Therefore either the names now used under the 'class system' are of unknown original sense; or, human marriage was, from the, first, so far 'individual' that our ideas of father, mother, brother, sister, son, daughter, could arise and could find expression in terms that still survive, say, among the Urabunna or other Australians. But while tribal customary laws as to classes, totems, generations, marriage rules, and many other social duties were being evolved; some of the ancient names for father, son, brother, sister, were perhaps taken up and applied to each of the large sets of persons whose customary legal status was now (as groups coalesced into large tribes) on the level of actual fathers, sons, brothers, sisters, and the rest. Obviously, in a primitive group of a male senior, his female mates and children, there could not exist (other groups being, on my theory, strange or hostile) large sets of persons occupying a common legal status, as in modern tribes. The existence of such sets of persons is the result of the later and tribal society, of society in which many groups are reconciled and united in a local tribe. Only in such a tribe, which cannot be primitive, is the classificatory system of naming sets of people necessary. It is only in tribal law that the grades of customary status answering to all the many terms can exist, and tribes with their laws cannot be primitive. Most names for the various grades, therefore, are later than Mr. Darwin's hypothetical stage of small and perhaps hostile groups; they were, in a few cases, perhaps originally names for such relationships as our own father, mother, son, brother, &c., but in the evolution of tribal customary law, such names have been extended

82

out of their family, or fire-circle, into their tribal significance, out of recognised kinship, or close contiguity, into terms including all who have the same status, rights, and duties.

SUPPOSED SURVIVALS OF GROUP MARRIAGE

If our suggestion as to the origin and significance of the 'classificatory terms of relationship' be plausible, then the theory of a pristine past of 'communal' or of 'group marriage' will lose what Mr. Darwin deemed the chief evidence in its favour, the evidence from terms of relationship. But there remains the evidence from 'survivals,' in institutions. For example, among the Urabunna, women of a certain seniority, totem and 'phratry' are Nupa to men of the relative status among males. They are the men's potential wives. In actual practice each individual man has one or perhaps two of these Nupa women who are specially attached to himself, and live in his camp. They are his wives. But each man has also, or many men have, other women of the Nupa set, who by an allotment, which the elders arrange, are his Piraungaru, He is, that is to say, their 'second master,' after their husbands. This is a kind of Cicisbeism, recognised and regulated by customary law, and sanctioned by a definite ceremony. Messrs. Spencer and Gillen therefore say 'individual marriage does not exist, either in name or in practice, among the Urabunna tribe.' Their idea appears to be that once every man was the husband of every Nupa woman who was accessible, and that the Piraungaru arrangement is a nascent restriction upon, or survival of, this communal marriage. It is admitted that a man may now try to prevent his wife from having sexual relations with her Piraungaru man, just as an Italian of the eighteenth century might have done in the case of his wife's Cicisbeo. 'But this leads to a fight, and the husband is looked upon as churlish.' The Italian husband would have undergone the same reproach, yet he lived in a society which in theory, and as Christian, insisted on individual marriage.

The question arises, is the Piraungaru arrangement a modified survival of communal marriage, or is it a mere chartered libertinism in customary practice, and not a 'rudimentary survival'? It is certainly found among the tribes most tenacious of archaic institutions. Mr. Crawley thinks, however, and, under correction, I agree with him, that the Piraungaru system is no survival, and that it 'has never been more fully developed than it is now.'[159]

---

[159] Spencer and Gillen, pp. 62-64. Mystic Rose, pp. 477-478.

PIRAUŊGARU AND PIRAURA

As to this Piraungaru affair, as usual we need, and do not get, the help of philology. What does the word 'Piraungaru' literally mean? Among the Dieri the Piraungaru custom prevails, and the persons affected by it are called Piraura—the resemblance to Piraungaru is striking. Now Mr. Howitt tells us that the Headman of the Dieri is called Pinaru, from pina, 'great,' but he also calls these Headmen Piraurus, the same title as he gives to the men and women allotted to each other on the system of native Cicisbeism.[160]

Clearly there is here either a misprint, or a curious fact. Either the Headmen are Pinarus, not Piraurus, or Headmen and supplementary wives and husbands have one and the same title! One great Headman was Jalina Pira murana. Is 'great' pina or pira? If Australia does not produce an adequate philologist in the native tongues, who will specially study these matters, it will be a heavy blow to the research into native institutions.

It is worth observing that the Dieri Piraura are 'permitted new marital privileges at the ceremony of circumcision.' Now license amidst the large assemblies brought together from all quarters on such occasions (in some places even transgressing the sacred rules of totem, phratry, and close relationship in our sense) is merely part of that periodical general 'burst' which survived in the Persian Sacæa and Roman Saturnalia. Many examples may be found in Mr. Frazer's 'Golden Bough.' Every kind of law is, at these 'bursts,' deliberately violated. Perhaps, then, the due selection of Piraura, by the Dieri seniors, is really rather a restriction of Saturnalian license than a relaxation of marriage laws, or a survival of communal marriage. That the license of the Saturnalia was a return to primitive ways was a Roman theory. For Australia it is the theory of the Arunta themselves.[161] The adjacent Urabunna have the same Piraura usages, and what looks very like a form of the same word, Piraura, Piraungaru. The relations thereby indicated exist, when occasion serves, after the season of license.

A wife, at marriage, is subjected to a disgraceful ordeal (modern ideas will break in), which I take, as Mr. Crawley does, to be a mere initiation (due to a well-defined superstition) into the life matrimonial.[162] Meanwhile, though a definite and disgusting set of proceedings forms the Urabunna marriage ceremonial, I am not aware that the same doings precede and sanction the establishment

---

[160] On the Organisation of Australian Tribes, pp. 107, 108.
[161] Spencer and Gillen, p. 97.
[162] Ibid. pp. 92-96. The Mystic Rose, pp. 479, 480.

84

of the Piraungaru or Piraura relation, which, if not, is no marriage at all. Thus, so far as our information goes, and with all deference to the great Australian authorities, I do not see that the evidence for a past stage of communal or of group marriage is such as compels our assent. On the other hand, as has been shown, the theory of communal marriage forces all its advocates, unwillingly or unconsciously, into the other theory of a primeval moral and social reformatory movement, deliberately undertaken, perhaps under direct divine inspiration, for what other motive could exist? The economical and biological difficulties which also beset that hypothesis have been sufficiently explained, and Mr. Darwin has dwelt on the psychological difficulty, the sexual jealousy of the primitive male. These objections, at least, do not hamper the hypothesis or conjecture, which we have ventured to submit as an alternative system. As a proof of survival of communal or group marriage, Mr. Fison quotes Mr. Lance: 'If a Kubbi meets a strange Ippatha' (female), 'they address each other as spouse.' (They belong to intermarrying phratries.) 'A Kubbi thus meeting an Ippatha, though she were of another tribe, would treat her as his wife, and his right to do so would be recognised by her tribe.' His right, as far as phratry prohibitions go, would certainly be recognised, but how her husband, if she had one, would view the transaction is another question. The morality is that of the Scottish ballads, in which such bonnes fortunes are frequent, and the frail pair only ask questions— afterwards. In the ballad of The Bonny Hind, in the Kalewala, and elsewhere, the answers prove that the pair are brother and sister. Suicide follows, but it does not follow that communal or group marriage prevailed in Scotland, or in Finland.

GROWTH OF SOCIAL RULES IN THE TRIBE

It is probable that the rules now defining the privileges, prohibitions, and duties of sets of people, rules interwoven now with those of 'class' and totem, have been gradually evolved in the wear and tear of ages. Tribes which hold such large and protracted assemblies, or palavers, as the Arunta of to-day, discuss and debate common affairs with all the diffuseness of our Parliament at Westminster. It is not to be supposed that tribal peace existed over hundreds of square miles of country, and that the group representatives, so to speak, flocked in from far-off regions, to parliament, in the ages when the pristine rules of exogamy were evolved. We might as wisely imagine that, in the beginning of Totemism, groups travelled to a tribal folk-mote, and arranged the

85

details of a kind of magical co-operative society to preserve and increase the foodstuffs of the tribe. In ages really pristine the tribal peace and union cannot have arisen; deliberate legislation for a vast scattered tribal community could not have entered into men's dreams. No such community could have existed. But the tribes of to-day, and notably the Arunta, being remote from truly primitive conditions, do hold prolonged assemblies, and work at public problems, so very remote from the primitive are they.

The Arunta, in their pseudo-historic legends, throw back upon the past the reflection of their actual estate, and ascribe the rule which practically limits marriage within the generation to a leader of the Thurathwerta group, living near what is called, by Europeans, Glen Helen, in the Macdonnell range. He was backed by the Emu people of four widely separated localities.[163] One is not, however, to suppose that, at some witan of the tribes, names indicative of generations, and of their respective rights, were suddenly invented and dealt out by 'the legislator,' any more than that totems were thus invented and dealt out. As Mr. Atkinson remarks (Chapter VIII.): 'Gradually each generation ... would, qua generation, come to be a distinctly defined class, with certain separate rights and obligations. In this simple classification of the connected persons, we see the origin of the classificatory system itself' (as far as generations are concerned), 'as an institution.... The classificatory system evolves itself merely as the result of a desire to define certain rights, and the division by generations was the most natural and feasible for the purpose.... Thus we find a desire for distinction, as regards rights in sexual union, to be the genetic cause of the classificatory system, both as concerns the generation and its component members.'

The marriage rules prevalent, with many variations, among the people least advanced in material culture, the Australians, are thus seen, on the whole, to be based (1) on totem rules (in which, with Dr. Durkheim, we include the 'primary classes' or 'phratries'), and (2) on the distinction of generations. It is clear, from the case of the Arunta and other tribes, that the rule of counting on the spindle side may break down, male descent being substituted, in times excessively rude; while again, as in the Pictish Royal House, it may elsewhere last into a stage relatively civilised. All manners of conditions and superstitions may affect and alter the course of social development in various places.

---

[163] J. A. I. vol. xviii., no. 3, pp. 245-272.

GROUP MARRIAGE AND MR. TYLOR's STATISTICS

In 1899, Mr. Tylor published a sketch of 'A Method of Investigating the Development of Institutions, applied to Laws of Marriage and Descent.'[164] He had catalogued the usages of 350 peoples, and examined (1) the rule of avoidance between husbands and wives' relations, and vice versa. (2) The naming of husband (or wife) after their children; as Odysseus says, 'May I no longer be called the father of Telemachus.' (3) The nature of inheritance in widows. (4) The Couvade in which the husband pretends to lie in, while his wife is really doing so. (5) The custom of capture. (6) Exogamy and the classificatory system. Mr. Tylor was led to believe that, so far as the statistical evidence goes, the husband first lived with the wife's family (A); next, after a residence with the wife's family, went back to his own home (B); last, (C) took the wife at once to his own home. (Husband to Wife. Removal. Wife to Husband.)

Now statistics are rather vague evidence without full knowledge of the social concomitants in each case. In what exact stage of culture, in each instance, does the husband go to live with the wife's relations? We have not this information. But if this be really the earliest stage, how is it compatible with group marriage? If a man is husband to 'a thousand miles of wives,' how can he go and live with the relations of all his wives? Even within his actual region of wandering, how can he do this? Nor, perhaps, can he bring all his wives to live with the relations of each of them in turn?

Either there was no group marriage, or it did not exist when, on the hypothesis, the husband, in the earliest stage, habitually resided with his wife's relations. Again, take the maternal and paternal systems, the reckoning in the female or male line, the female line, as we hold with Mr. Tylor, being the earlier. If so, on Mr. Tylor's hypothesis, it ought to arise in his first epoch, when husband goes to live with wife's people. 'The lines' (of a diagram) 'show the institutions of female descent, avuncular authority, &c. arising in the stage of residence on the female side, and extending into the stages of removal and residence on the male side.'

Now we have tried to explain the reckoning in the female line, by the differentiation, in the supposed original local totem group, of the captive women, each retaining, and handing on to her children, the name of her own totem group, this bequest of the totem name continuing into the tribal state of peaceful betrothals. But Mr. Tylor's theory of the first stage (husband goes to live with wife),

---

[164] Spencer and Gillen, pp. 420-421.

implies a peaceful state, and groups not hostile. For the reasons given, early hostility and sexual jealousy, I am unable to hold that, in the beginning, husbands always joined their wives' groups. It seems, granting hostility and jealousy, to be impossible. A Malay example of polygamy plus residence with wives' relations, proves nothing for primitive man. Therefore we need to know the exact stage of culture of the peoples among whom the husbands go as subdued hangers-on, 'not recognised,' into the wife's family. Are these people all precisely primitive? The 'husband to wife' stage implies peaceful relations. These were produced, on my theory, by the arrangement of the phratries. When these are once constituted, the husband may go to live with the wife's family as much as he pleases. But I fail to see how he could have done so 'in the beginning.' Moreover I am disinclined to suppose that exogamy was instituted for the purpose of strengthening a group by matrimonial alliances.

Bella gerant alii, tu, felix Austria, nube!

Exogamy has this effect, but it was not devised purposely to produce this effect.

I may casually remark that Mr. Tylor mentions an Assineboin case in which the husband enters 'his lodge,' where his father and mother-in-law 'shirk' or avoid him. But, in the next page but one, the lodge is described, not as the husband's, but as that of the father and mother-in-law.[165] Whose lodge was it really? Was the husband staying with his wife's family, or were the old people on a visit to their married daughter?

Among several Australian tribes, a feigned form of capture precedes marriage.[166] Is this a survival of actual capture in the stage of hostility, the pre-tribal stage? Or is it the result of girlish modesty in the bride?

Note: Mr. Morgan's 'Reformatory Movement': It is proper to note that, in his preface to Kamilaroi and Kurnai (p. 5), Mr. Morgan wrote, 'it is not supposable that savages design, consciously, reformatory movements in the strict sense.' For his theory cannot escape the conclusion that, in fact, they did.

---

[165] Op. cit. pp. 246-248.
[166] Howitt, Organisation of Australian Tribes.

CHAPTER VI

THE CHANGE OF CLASS AMONG THE NEW GENERATION

We have hitherto, for the sake of lucidity, spoken chiefly of two 'primary classes' ('phratries'), such as the Kirarawa and Matthurie of the Urabunna. But among the Arunta, and many other tribes, there are four or even eight such 'classes.' The reader may refer to the extract from Mr. Mathews's description (p. 39).

Each of these classes roughly corresponds to a different generation of the tribe. But, with female descent, each child belongs to the class to which its mother does not belong. The classes, that is, alter with each generation. What is the cause of this curious rule? One generation is A, its children are B, its grandchildren are A again.

Here we meet the explanation of Herr Cunow, which may as well be given in summary.

THE SYSTEM OF HERR CUNOW

The theory of Herr Cunow[167] is in the first place opposed to the systems of all who regard the 'phratries' as divisions made in an original group, or horde, for purposes of exogamy. I have not observed that any of our writers have noticed the book of Herr Cunow. In his opinion, as was said earlier, authors err in confusing 'phratries' with 'classes:' 'a phratry is not a class, and a class is not a phratry; these two sorts of bodies have been developed out of different antecedents, and have different tendencies. The two "primary divisions," say Kroki and Kumite, are phratries, but are not classes in the same sense as the Ippai and Kumbo, Murri and Kubbi classes of the Kamilaroi' (p. 24).

Herr Cunow regards the 'classes' as in origin earlier[168] than the divisions of totem kin, or the 'phratry' divisions, and thinks that the 'classes' were originally non-intermarrying divisions based on seniority. They were devised or developed, not to prevent marriage

---

[167] Die Verwandtschafts-Organisationen der Australneger. Diek, Stuttgart, 1894.
[168] This can hardly be, as the most backward tribes have phratries and totems, but no 'classes.'

between near kin, but between persons of different generations, or rather degrees of seniority. This is proved, he thinks, by the etymology of some of the names of the classes (about which we need much fuller information). Thus the word Kubbi (Kamilaroi), already cited as a class name, is derived, he says, from Kubbura, 'young, new,' and originally designates a youth who has passed the initiatory ceremonies. Ridley's vocabulary of the Kamilaroi tongue is the source for this fact. Kumbo, another class name, is the Kombia or Kumbia, of the tribes on the Lower Murray river, and means 'great,' that is, 'old.' On the Lower Darling, the word is gumboka, Kumbuka; compare Kumba, Kumbera, 'old woman,' Kumbeja, 'father.' 'Great' and 'old,' 'little' and 'young,' are equivalent in sense. Bonda, a class name of the Kabi, means 'new' or 'young,' and the class-name Darawang, or Tarawang, is the Kabi word darami, 'little,' or 'young.' Obu, a class name, is the Queensland jabu, jobu, jabbo bobu, 'father.'

Thus the class names, Herr Cunow holds, originally indicate divisions of youth and age in the 'horde,' by which term Herr Cunow understands a local set of from forty to sixty people, a local aggregate of several such 'hordes' being a 'tribe' (pp. 25-28). The fact of Australian attention to degrees of seniority is demonstrated by the stages of initiation, and by the various dues, of food gifts and so on, paid by the juniors to the seniors of the tribe: by the food which persons of different status in seniority may eat, and so forth. Indeed Dr. Roth has regarded the 'classes' as originally evolved to regulate the distribution of the food supply, and such regulations would, I think, be elements among other regulations of matrimonial and other rights, dependent on seniority. 'What a man may eat at one stage is at another stage forbidden, and vice versa.'[169]

The 'horde,' then, in Herr Cunow's opinion, was primarily divided into non-intermarrying persons of three stages of seniority. This is the original organisation, that of totem kindreds being later, in Herr Cunow's theory, which is not ours (pp. 36, 37). The word 'father' does not, in the Australian dialects, at first, signify what we mean by the word, but merely 'senior;' and 'mother' is a term of the same meaning. 'Father' and 'mother' with all of their seniority are 'the big ones;' children are 'the little ones.' These terms become 'class' names.

An example is taken from Mr. Bridgman, superintendent of the tribes at Port Mackay. These have two 'phratries,' Yungaru and Wutaru (totemic names), and four 'classes,' Gurgela, Bembia,

---

[169] Eyre, Journals, ii. 293-295. Cunow, p. 33, note 2. Bulmer, in Brough Smyth, i. 235. Roth, Ethnological Studies, pp. 69, 70, Brisbane, 1897.

Wungo, and Kubaru.[170] The terms for family relations are not understood in our sense. Mr. Bridgman had a name and status in the tribe. His name was Gunurra; his phratry was Yungaru, his class was Bembia, and his children, if he had any, were Wutaru (by phratry), Kubaru (by class). If a girl came by, and Mr. Bridgman asked who she was, and if she was Kubaruan, he was told 'she is your daughter.' This 'daughter' is a young woman of the class to which Mr. Bridgman's daughters, if he had any, would belong.

Herr Cunow's theory, then, starts from the 'horde,' divided into not intermarrying degrees of seniority. That such hordes, not separate family groups, were the initial stage of society, he is persuaded.[171] He rejects Morgan's theory of communal marriage.[172] Next, he thinks, arose objections to brother and sister and other near akin marriages (why we are not told), and a man would thus be driven to seek a wife out of his own horde. Why was this? Herr Cunow merely refers to the Dieri tradition already cited; evils followed on kindred marriages, and were perceived and, by divine decree, were reformed.[173] That such evils did arise and were perceived, and being perceived were reformed, by very low savages, is to the highest degree improbable. However it came about (we suggest by dint of reflection on the totem and phratry restrictions), there is now an objection to intermarriage between persons 'of the same flesh.' How this arose does not seem to be a question that Herr Cunow chooses to dogmatise upon.

The horde now develops itself into a group of kin, of which the members regard each other as 'too nearly related by blood,' to intermarry. 'As a mark of these groups of kin they later take different beast or plant names, usually from such species as exist in their districts. No reverence would originally be paid to the totem animal;' the Narrinyeri eat it without scruple,[174] like any other; the totem name is originally a name of a genossenschaft; a comradeship, the Narrinyeri word for totem, 'Ngaitje,' is equivalent to 'friend.'

All this is rather vague. Why did groups of comrades or of recognised kin take plant and animal names? Why did they forbid intermarriage? What was the origin of the objection to marriage between blood kindred? It does not arise out of 'moral ideas,' nor

---

[170] Brough Smyth, i. 91.

[171] Pp. 122-124, and note 1, an argument against Westermarck.

[172] Pp. 127-128.

[173] Gason, The Dicyrie Tribe (1894), p. 13. Kam. and Kur. p. 25. Cunow, pp. 109-110, 130-132.

[174] Cf. Cunow, p. 82. So, too, the Euahlayi.

out of 'wife-capture,'[175] and Herr Cunow speaks neither of 'sexual taboo,' nor of 'sexual jealousy,' while the theory of 'personal totems' become hereditary, or of magical co-operation in totem breeding, is not mentioned; indeed, when Herr Cunow wrote (1894), the magical theory was unborn. The hordes merely developed into groups of comrades or of kin, as such not intermarrying among themselves, and marking themselves for no assigned reason, with plant or animal names: reverence of the totem came later.

'Still later than the totem association the phratry seems to arise,' and the phratries are described as allied local totem groups. This is my own opinion, but by 'local totem group,' I here mean (as already explained), the original local totem group, with the other totems which had become its elements, through exogamy, and female descent. Herr Cunow, if I follow him, means on the other hand a local totem group of the kind which now results among the Arunta from reckoning descent in the male line. 'The forbidding of marriage extended beyond the local group, passing into the neighbouring hordes, till at length morality enjoined the obtaining of wives from remoter districts. Hence was developed a come-and-go of marriage between two out of several larger local totems, and these larger local communities are the original types of the Australian phratries. Suppose that the hordes of the Kurnai had gradually developed themselves into local totem groups like those of the Narrinyeri, and ... that it became the rule for the Brataulong to take their wives from their south-western neighbours, the Kulin, and vice versa, till the two groups waxed into a great community, and we have the probable development' (of the 'phratries') 'before us.' The groups 'Brataulong' and 'Kulin' would now be a great community of two intermarrying phratries.

All this implies, I think, a more advanced society, and larger communities, than we can easily conceive to have existed in the distant past when phratries arose. Moreover Herr Cunow, as we shall see, takes descent, even at this primitive period, to have been reckoned in the male line. Again, we have observed that phratry names, when they can be translated, are usually totemic, an opinion expressed by Mr. Fison and Mr. Howitt. The same sort of totemic names marks Red Indian phratries. Granting male kinship, the phratries of Herr Cunow's hypothesis might well have totem names, but he tries to show that phratry names are usually local; he gives seven cases out of which only two names of phratries are totemic.[176] But he offers no authority for his assertion that the other five names

---

[175] Cunow, p. 130.
[176] Pp. 133-134.

are non-totemic (Eigennahme) and Yungaru and Wutaru, represented by him as non-totemic, are really totem names.

We know that as a result of reckoning in the male line local or district names tend to supersede totem names, and large local totem groups thus arise, a feature of the decay, not of the dawn, of Totemism. My own hypothesis, on the other hand, shows why phratry names are totemic. Herr Cunow concludes 'the phratry is originally nothing but an exogamous local group composed of several hordes.' Like Mr. Daniel McLennan, Herr Cunow quotes the legend of the wars of Eagle-Hawk and Crow, which ended in the establishment of the intermarrying phratries of Crow and Eagle-Hawk.[177] Herr Cunow's theory of phratries appears to me to find, in the remotest past, the most recent institutions of the Australians, and to confuse the primitive local totem group with the local totem-group later developed out of reckoning descent in the male line. He throws back into the distant past the large modern associations, which could not exist in times really primitive. He makes the hordes develope themselves into totem kins, in place of being, originally (as in my system), small associations united by contiguity, and receiving totem names from without.[178] He makes reckoning in the female line later than reckoning in the male line—the Narrinyeri reckoning in the male line (p. 84)—and perhaps this method, he thinks, is a result of ignorance of fatherhood, consequent on the Piraungaru custom (p. 135). Unluckily we find reckoning descent in the female line among many races, the Red Indians for example, where the Piraungaru custom is unknown. The priority of male to female descent is not admitted as a rule, by Mr. Tylor or any other English authorities.

Where I can agree with Herr Cunow is on the point that the two 'primary divisions' are the result, probably, of amalgamation, not of bisection for purposes of exogamy. Where we differ is as to the character of the communities that, by alliance and connubium, became 'primary divisions' or 'phratries.' On his system the communities were large, holding great districts. On mine, they were ancient local totem groups, whose members, through exogamy and female descent, were really of various totems. In a note (p. 139) Herr Cunow shows that he might easily have arrived at my conclusion, but, while allowing that alien brides brought the totem names of their own kins into each original totem group, he says that the men of that group still 'belonged to the totem identified with

---

[177] Brough Smyth, i. 423. Cunow, p. 134. Studies in Ancient History, second series, ut supra.
[178] See 'The Origin of Totemism.'

that horde.' This is the result of his belief that reckoning descent in the female line is 'an innovation.' His 'horde' is originally endogamous; then, we know not well why, is exogamous (p. 137). Those who do not believe that men originally lived in 'hordes,' and hold that, through jealousy and other causes, their little primary sets were, or tended to be, exogamous from the first, cannot agree with Herr Cunow. On the other hand, they may incline to accept his theory that, as the Australian terms of relationship indicate often status, not relationship in our sense, they do not help to prove a past of consanguine and communal marriage.

CLASSES AGAIN

To return to the classes, Dr. Durkheim opposes Herr Cunow's theory that they indicated originally degrees of seniority. He takes no notice, however, of Herr Cunow's argument from etymology, and the original meanings of the class names, 'Young' and 'Old.' He argues that, on Herr Cunow's system, each individual would, in lapse of time, move from young to old, and so ought to change his class name, and move into another class. Herr Cunow answers that, if this occurred, the object of the class names, practically to prevent young and old intermarrying, would have been defeated. But, as matters exist, a grandfather may marry a girl who might be his grand-daughter. He is A, his children are B, but their children are A again. He is Kubbi, he marries Ippatha, her children are Buta, their children are Ippatha, and the venerable Kubbi may marry a very juvenile Ippatha.

Possibly the institution grew up among people who did not look so far forward, who 'took short views.' It is certain that, if the object of the classes was to stop marriages between young and old, it is a failure. 'The old men marry young wives at present,' says Mr. Mathews. If so, Herr Cunow may be right. Dr. Durkheim offers a theory. But his theory takes for granted, as we saw, that the two 'phratries,' originally, were only two totem groups, containing within them no members of other totem kins. 'They were not yet subdivided' into other totem kins. But I have tried to show that there was no such 'subdivision' into 'secondary clans' or totem kins. Dr. Durkheim regards these totem kins as colonies split off from the two original totem groups which became phratries.[179] My reasons against accepting this position have already been given. This being the case, it is unnecessary to unfold Dr. Durkheim's theory of the

---

[179] Cf. p. 83.

origin of the classes. Probably that of Herr Cunow comes nearest to the truth.

Mr. Mathews offers another solution of the problem. 'Phratry' Dilbi, for example, has 'classes' Murri and Kubbi, while the linked phratry, Kupathin, has classes Ippai and Kumbo. 'It is possible that the group Dilbi was divided into (female) Matha and Kubbitha to distinguish the mothers from the daughters, and that the terms Murri and Kubbi were adopted to provide names for the uncles and nephews of their respective generations.' Thus we return to distinction of generations. In any case the 'classes' 'have the effect of preventing consanguineous marriages, by furnishing an easy test of relationship when the tribe has become so numerous or widespread that kinship could not otherwise be well determined.[180] Later (p. 168) Mr. Matthews writes, 'The mother of a man's wife, and also his daughters, belong to the same section' ('class'), 'and therefore his marriage with that section is prohibited.' That is, he cannot marry out of his generation above or below, as indicated by 'class' names. 'Neither can he marry into the section to which his mother belongs, although a woman might be found in either case, who was in no way connected with him.' In short, as far as the names rudely indicate the generation above, and the generation below a man, he cannot marry into these classes. But, as old men do marry young wives, the apparent intention of the rules is to some extent frustrated. We can say no more, till we are told what the class names mean in a literal sense. Does nobody inquire into this essential question?

As if to accentuate the problems raised by the change of 'class' names in each generation, Mr. Matthews has discovered that when a man may marry a woman of his own 'phratry,' but out of a set of totems not his own, the totems of his children by her alter as the class names do. 'The children take the totem name,' not of their mother, but of their maternal grandmother. 'One totem is the mother of another totem.'[181] This is an unusual phenomenon, and looks like the effort of a desperate ingenuity.

The class system exists among the Arunta, with male descent. One moiety of the southern part of the tribe consists of Panunga and Bulthara, linked classes, calling themselves Nakrakia; the other moiety is of Purula and Kumara, calling themselves Mulganuka. A Bulthara man of the first moiety can only marry a Kumara woman, of the second moiety: a Purula man marries a Panunga woman only. The children of a Bulthara man's union with a Kumara woman take neither the Bulthara nor Kumara name, but are called Panunga,

---

[180] Proc. Roy. Soc. N.S.W. xxxi. 161.
[181] Op. cit. pp. 172-175.

while the children of a Purula man and a Panunga woman are Kumara: of a Panunga man and a Purula woman, Bulthara; of a Kumara man and a Bulthara woman, Purula.

That is to say, the Arunta reckoning in the male line, a man's children do not take his 'class' name but the name of the 'class' linked to his, and forming, with his, one division of the tribe. Further each of these four divisions consists of two moieties, and a Panunga man, though he can marry a Purula woman, must choose her out of the proper moiety of the Purula division. These moieties of each division, among the Northern Arunta, have names; Uknaria, Appungerta, Umbitchaṇa, Ungalla, and the children of each marriage fall under these names.

This restricts a man to only an eighth of the women of his generation, but, on the other hand, among the Arunta, the totem prohibition no longer exists: the totems are not restricted to one or another class, but skip among them, as we have shown in the section on the Arunta. The eight class system, perhaps the four class system, may be regarded as later and conscious modifications of the old phratry and totem rules, which, on my hypothesis, had no conscious moral origin.

CHAPTER VII

THEORIES OF LORD AVEBURY

The opinions of Lord Avebury (Sir John Lubbock) are to be collected from the sixth edition (1902) of his Origin of Civilisation. First published in 1870, this was a pioneer work of great value and importance. Perhaps the vast amount of new information and of new speculation which has accrued since 1870 might almost make us wish that Lord Avebury had found time to re-write his early book. But he 'sees no reason to change in any essential respects the opinions originally expressed,' and merely adds a few references to such recent researches as those of Messrs. Spencer and Gillen. Therefore we must not look to Lord Avebury for much new light on the origin of the Australian 'classes' or 'primary divisions,' or 'phratries,' and on their relations to the totem kindreds within them.

LORD AVEBURY ON TOTEMISM

Our author (p. 217) regards Totemism as synonymous with Nature-Worship. He speaks of 'Nature-Worship or Totemism, in which natural objects, trees, lakes, stones, animals, &c. are worshipped.' I am not acquainted (unless it be in early Peru) with any totem kin whose totem is a lake; and totems, very often, are not 'worshipped' at all. Nature-Worship, again, may exist where there is no Totemism, and Totemism where there is no Nature-Worship, indeed where, as among the Arunta, there is, strictly speaking, no worship, as far as we are informed.

Again (p. 351), 'Totemism' (as opposed to fetichism), 'is a deification of classes.' But the term 'deification' implies the possession, by the deifiers, of the conception of Deity; of gods, or of a god. The Australians have totems, but, according to Lord Avebury, have no notion of a god or gods. They 'possess merely certain vague ideas as to the existence of evil spirits, and a general dread of witchcraft' (p. 338). It is not clear, then, how they can 'deify' classes of things, if they have no notion of deity. 'They do not believe in the existence of a true Deity' (with a capital D), says Lord Avebury, without defining what 'a true Deity' is: and, contrary to the evidence of Mr. Howitt and many others, he denies that 'morality is in any way connected with their religion, if such it can be called' (p. 338).

The authority cited is of 1859,[182] and is contradicted, for example, by Mr. Howitt (1880-1890), who is not here quoted. It is clear that Australian totems cannot result from the 'deification of classes,' if the Australians have no conception of Deity, whether 'true' or not so true.

Lord Avebury remarks, 'True, myths do not occur among the lowest races' (p. 355), whereas, with many others, myths of the origin of Totemism do notably occur, as we have shown, among perhaps all totemistic races. Perhaps we should read, deleting the comma, 'true myths do not occur among the lowest races,' when the question as to what a 'true myth' is again arises, as in the case of 'a true Deity.' Perhaps we must suppose that by 'a true myth,' or a 'true Deity,' Lord Avebury implies a Deity or a myth in accordance with his own conception of either.

---

[182] 'Report of the Committee of the Legislative Council on Aborigines.' Victoria, pp. 9, 69, 77.

LORD AVEBURY ON THE ORIGIN OF TOTEMISM

'The worship of animals,' says our author (p. 275), 'is susceptible of a very simple explanation, and perhaps, as I have ventured to suggest,[183] may have originated from the practice of naming, first individuals, and then their families, after particular animals. A family, for instance, which was called after the bear, would come to look on that animal first with interest, then with respect, and at length with a sort of awe.' If by 'individuals,' male individuals are intended, this theory is open to the objection that Lord Avebury regards descent in the female as earlier than descent in the male line (p. 164), while 'families' with enduring relations to their founders, can hardly yet have been consciously envisaged, by his theory, at so very rudimentary a stage. Moreover, we try to show that totem names were, originally, group names, and were not derived from the personal names of individuals, an opinion in which Mr. Haddon concurs. Lord Avebury's theory is, apparently, that of Mr. Herbert Spencer, minus the supposed worship of the ghost of the male ancestor and founder of the family.

COMMUNAL MARRIAGE

Lord Avebury assumes, as a working hypothesis, that 'the communal marriage system ... represents the primitive and earliest social condition of man....' (p. 102). The objections to this hypothesis we have stated, though, of course, historic certainty cannot be attained.

Lord Avebury, assuming 'communal marriage' as the Primitive stage, holds that it 'was gradually superseded by individual marriage founded on capture, and that this led firstly to exogamy, and then to female infanticide; thus reversing Mr. McLennan's order of sequence' (p. 108). 'Originally no man could appropriate a woman of his own tribe exclusively to himself ... without infringing tribal rights, but, on the other hand, if a man captured a woman belonging to another tribe, he thereby acquired an individual and peculiar right to her, and she became his exclusively, no one else having any claim or property in her' (p. 110). (I here italicise 'tribe' and 'tribal.' Lord Avebury intends, I think, a woman of the same 'fire-circle' (p. 188), not a woman of the tribe understood as a large and inevitably not primitive local aggregate of

---

[183] Prehistoric Times, p. 598.

friendly groups of different totems, such as the Arunta, Narrinyeri, Pawnees, and so forth.)

In brief, men would desire to appropriate to themselves some woman, at first from beyond their own 'tribe.' This they could only do by capture. Their individual right in her would be modified by the disgusting license of the bridal night, which Lord Avebury regards as 'compensation' to the other males of the 'tribe' (pp. 138, 557-560). That license I would rather explain as Mr. Crawley does: the topic does not need to be insisted on at length in this place. Lord Avebury, at all events, supposes that a form of capture finally came to be applied, with results in individual marriage, to women of the same 'tribe' (p. 111). But if we have 'complete and conclusive evidence that in large portions of Australia every man had the privilege of a husband over every woman not belonging to his own gens; sharing, of course, these privileges with every other man belonging to the same class or gens as himself' (p. 112), I fail to see that a man gained anything by enduring the trouble and risk of capturing a bride all to himself. Before the capture she had been, it seems, the common spoil of the males of her 'tribe;' when captured she was the common spoil of her captor's 'class or gens'—though a 'class' and a gens are not, I think, identical, but much the reverse.

The rather promiscuous use of terms for different kinds of human communities affected all the pioneer works on primitive society, and, indeed, still perplexes our speculations. Thus Lord Avebury suggests (p. 119) the case of four exogamous neighbouring 'tribes,' with kinship traced through women. 'After a certain time the result would be that each tribe would consist of four septs or 'clans' (totem kins?), 'representing the four original tribes, and hence we should find communities in which each tribe is divided into clans, and a man must always marry a woman of a different clan.'

We do not, perhaps, know any exogamous tribes in our sense of 'tribe;' a Dieri is not obliged to marry out of the Dieri, or an Urabunna out of the Urabunna. By 'tribe' here, it seems probable that Lord Avebury intends not a large local aggregate, but 'a very small community,' for he writes 'we have seen that, under the custom of communal marriage, a child was regarded as related to the tribe, but not specially to any particular father or mother. Such a 'state of things, indeed, is only possible in very small communities.' Now a tribe is a very large community. The members of such communities must have been poor observers if they did not discover the relation between a child and the woman who bore and, for several years, nursed it. But such 'tribes' are not tribes in the sense in which I use the word; they are rather 'groups of the same hearth.'

Now it is easy to see how small groups of the same hearth became exogamous, namely through sexual jealousy, and sexual tabu, which would oblige the young males to wander away, or to get wives by capture, practices resulting, under the tabu, in the sacred rule of exogamy. This, however, is not Lord Avebury's theory of the origin of exogamy.

Lord Avebury's theory does not become more distinct when he says, 'In Australia, where the same family names' (totem names?) 'are common almost over the whole continent, no man may marry a woman whose family name' (totem name?) 'is the same as his own' (here the Arunta are an exception) 'and who belongs therefore to the same tribe' (p. 144). But surely, if the 'family names' are 'common almost over the whole continent,' a woman may well have the same 'family name' (say Emu) as a man, and yet need not be of his tribe. An Arunta Emu man and a Dieri Emu woman would have the same 'family name' (totem name), but would not, therefore, 'belong to the same tribe.' It even appears that Lord Avebury regards 'tribe' and 'clan' and 'family' as synonymous terms, for, in proof of the statement that people of the same 'family name' necessarily belong to the same 'tribe,' he quotes my late uncle, Mr. Gideon Scott Lang, 'No man can marry a woman of the same clan, though the parties be no way related according to our ideas.'[184] By 'clan' Mr. Lang here meant totem kin, and if Lord Avebury thinks 'clan' equivalent to 'tribe,' a 'tribe' must be a totem kin, which it is not; at least if we understand 'tribe' as a local aggregate of various totem kindreds.

These perplexities are caused by a vague terminology, and occurred naturally in a book of 1870, as they do in Mr. McLennan's own pioneer works. But in 1903 we must try to aim at closer and more exact distinctions and definitions, though we are still retarded and perplexed by the lack of truly scientific nomenclature. As far as I can perceive, Lord Avebury is apt to use 'family,' 'tribe,' 'clan,' and 'gens' as equivalents, while each of them, in various places, appears to be understood as denoting a totem kindred. Thus (p. 181) 'under a system of female descent combined with exogamy a man must marry out of his tribe,' where 'tribe' seems to mean 'totem kin.' Compare p. 187: 'another general rule, in America as elsewhere, is that no one may marry within his own clan or family,' where 'clan or family' like 'tribe' seems to mean 'totem kin.'

This use of terms makes it difficult for me to feel sure that I apprehend Lord Avebury's theory correctly. However I take it to be that, originally,'very small communities' ('tribes') lived in

---

[184] G. Scott Lang, The Aborigines of Australia, p. 10.

'communal marriage.' Nobody knew who was the son of what father or of what mother, though, in a very small community one would expect the senior vigorous male or males to prevent son-and-mother, or brother-and-sister unions, by force, out of natural jealousy. This was not done, but some males wanted wives to themselves in private property, and got them by capture, paying 'compensation' in the license of the bridal night. But a man might fall in love with a lass in his own 'tribe' ('very small community') and want to keep her to himself (p. 111). 'Hence would naturally arise a desire on the part of many to extend the right of capture, which originally had reference only to women of a different tribe, and to apply it to all those belonging to their own.' Is 'tribe' still used of 'a very small community,' or is it here employed in the now more prevalent and much wider sense? If not, is the 'capture' now a mere ceremonial formula? Apparently 'tribe,' now and here, does mean (as elsewhere it does not) a large local aggregate, for we are next told of 'the division of Australian tribes into classes or gentes' (though a 'class' is one thing and a gens, if totem kin is meant, is another thing), and of the '1,000 miles of wives,' who, by the theory, are not individual wives of individual men. Such wives, special rights in such wives, were acquired 'originally by right of capture.'

But, when men possessed marital privileges, each 'over every woman not belonging to his own gens; sharing, of course, these privileges with every other man belonging to the same gens or class as himself' (p. 112), where is the individual right acquired by capture? It seems that each man, besides his '1,000 miles of wives' 'has his own individual wife ... by right of capture.' Now the Urabunna have no such individual wives, if, like Lord Avebury, we accept the statement of Messrs. Spencer and Gillen (p. 63). But the Arunta have such individual wives. Here it seems necessary for Lord Avebury, if he agrees with these authors, to prove that the Arunta, unlike the Urabunna, do demonstrably acquire their individual wives by capture. But no such demonstration is produced. Till proof is offered I am unable to appreciate the force of Lord Avebury's reasoning, while like Mr. Crawley, I doubt whether individual marriage does not exist among the Urabunna, the Piraungaru license not being, I conceive, a true survival of communal marriage, but a peculiar institution.

LORD AVEBURY ON RELATIONSHIPS

Analysing Mr. Morgan's collection of names for relationships, Lord Avebury (p. 182) says, 'in fact the idea of relationship, like that of marriage, was founded, not upon duty, but upon power.' We try to suggest that the classificatory names for relationships are, to a great extent, expressive of status, seniority, and mutual duties and services in the community—these duties and services themselves being gradually established by power—the power of the seniors. Yet some terms analysed by Lord Avebury have, linguistically, other sources. 'Wife,' in Cree, is 'part of myself,' dimidium animæ meæ, these twain are one flesh. Obviously this pretty term does not spring from 'communal marriage.' In Chocta, 'husband' is 'he who leads me,'—again not communal, but indicating the old-fashioned theory of wifely obedience. ('He who kicks me' would suffice, in some civilised quarters.) 'Daughter-in-law,' in Delaware, is 'my cook,' indicating service; and 'husband' is 'my aid through life,' showing the advanced Homeric, or Christian, view of marriage (pp. 180-181). 'Father' and 'Mother' in many African, European and Asian, Non-Aryan, Oceanic, Australian, and, really in Aryan languages, also often in America, are 'the easiest sounds which a child can pronounce indicating father and mother' (pp. 442-449). If babes could distinguish father and mother, these relationships, one thinks, could not have been unknown to adults. They may be, and are, extended in usage, so as to embrace what we call uncles and aunts and seniors of the kin, but this, I try to argue, does not necessarily imply that fatherhood and motherhood, owing to communal marriage, were long unknown.

The result of Lord Avebury's analysis of Mr. Morgan's tables of terms is to prove progress in the discrimination of degrees of kin, though ancient sweeping terms occasionally survive among races fairly advanced out of savagery. 'Relationship is, at first, regarded as a matter, not of blood, but of tribal organisation' (p. 208). Here I agree that words or terms for what we call relationship often do seem to denote status, duty, service, and intermarriageableness in the community. But I do not think that the ties of blood are thereby proved to have been unknown. Maternity could not be doubtful, especially where the mother nursed her child for several years.

Lord Avebury adds, 'the terms for what we call relationships are, among the lower races of men, mere expressions for the results of marriage customs, and do not comprise the idea of relationship as we understand it' (p. 210).

For this reason, I think, we must avoid the fallacy of arguing as if the terms did denote 'relationship as we understand it,' when we wish to prove a past of communal marriage. The terms indicate, in Lord Avebury's words, 'the connection of individuals inter se, their duties to one another, their rights, and the descent of their property.'

This is precisely my own opinion, and for this very reason I do not hold that these terms arose in ignorance as to who was the mother, or even the father, of a child. All the duties and rights, as Lord Avebury says, 'are regulated more by the relation to the tribe than to the family'—in our sense of 'family.' But this, in my view, proves that the terms (in their present significance) are relatively late and advanced, for the institution of the Tribe (as I understand the word) implies the friendly combination of many totem kins, and of many 'fire-circles,' into the tribe, the large local aggregate. No such combination can have been truly 'primitive.' But we have seen that Lord Avebury seems to use 'tribe' in various places, as equivalent to 'family,' 'clan,' gens, and, apparently, to 'totem kin.' Quite possibly he means that the horde is prior to what I may call the 'fire-circle,' the 'very small community,' which, in places, he terms 'the tribe,' or so I understand him. If so, I cannot follow him here, as I am not inclined to think that truly primitive man lived in hordes of considerable numbers: the difficulties of supply, among other reasons, make the idea improbable.

If I have failed to understand Lord Avebury, perhaps his somewhat indeterminate terminology may plead my excuse.

CHAPTER VIII

THE ORIGIN OF TOTEM NAMES AND BELIEFS

Up to this point, we have treated of totems just as we find them in savage practice. We have seen that totem names are the titles of groups of kindred, real or imagined; they are derived from animals, plants, and other natural objects; they appear among tribes who reckon descent either on the sword or spindle side, and the totem name of each group is usually (but not in the case of the Arunta) one mark of the exogamous limit. None may marry a person of the same totem name. But, in company with this prohibition, is found a body of myths, superstitions, rites, magical practices, and artistic uses of the totem.[185] We have shown (Chapter II.) that we cannot move a step without a clear and consistent hypothesis of the origins of Totemism. This we now try to produce.

SACRED ANIMALS IN SAVAGE SOCIETY

Savages, both in their groups of kin, in their magical societies, or clubs, and privately, as individuals, are apt to regard certain beasts, plants, and so on, as the guardians of the group, of the society, and of the private person. To these animal guardians, whether of the individual, the society, or the group of kin, they show a certain amount of reverence and respect. That reverence naturally takes much the same forms—the inevitable forms—as of not killing or eating the animal, occasionally praying to it, or of burying dead representatives of the species, as may happen. But I am unaware that the savage ever calls his personal selected animal or plant, or the guardian animal of his magical society (except among the Arunta, where the totem groups are evolving into magical clubs), by the same term as he applies to the hereditary guardian of his group of kindred; his totem, as I use the word. If I am right, this distinction has been overlooked, or thought insignificant, by some modern inquirers. Major. Powell, the Director of the Ethnological Bureau at Washington, appears to apply, the word totem both to the chosen animal friend of the individual, and to that of the magical society in America, which includes men of various group totems.[186] He also applies it to the totem of the kin.

Mr. Frazer, too, writes of (1) The Clan Totem, (2) The Sex Totem (in Australia), (3) 'The Individual Totem, belonging to a single individual, and not passing to his descendants,' and even indicates that one savage may have five totems.[187] This third rule as to the non-hereditable character of 'the individual totem' has, since

---

[185] As to the word 'totem,' but little is certainly known. Its earliest occurrence in literature, to my knowledge, is in a work by J. Long (1791), Voyages and Travels of an Indian Interpreter. Long sojourned among the Algonquin branch of the North American Indians. He spells the word 'Totam,' and even speaks of 'Totamism.' Mr. Tylor has pointed out that Long in one place confuses the totem, the hereditary group name, and protective object, with what used to be called the manitu or 'medicine,' of each individual Indian, chosen by him, or her, after a fast, at puberty. Remarks on Totemism, 1898, pp. 139-40. Cf. infra, 135, note.
[186] Man, 1902, No. 75.
[187] Totemism, p. 2, 1887.

Mr. Frazer wrote in 1887, been found to admit of more exceptions than we then knew. In a few cases and places, the animal selected by, or for, the private individual, is found to descend to his or her children. In my opinion it is better, for the present at least, to speak of such protective animals of individuals, by the names which their savage protégés give to them in each case: nyarongs (Sarawak) 'bush-souls,' (Calabar) naguals, (Central America) manitus (?) as among the Algonquins, Yunbeai in some Australian dialects, and so forth.[188] I myself here use 'totem' only of the object which lends its name, hereditarily, to a group of kin.

PROPOSED RESTRICTION OF THE USE OF THE WORD 'TOTEM'

This restriction I make, not for the purpose of simplifying the problem of totemism by disregarding 'the individual totem,' 'the sex totem,' and so on, but because I understand that savages everywhere use one word for their hereditary kin totem, and other words for the plant or animal protectors of individuals, of magical societies, and so forth. The true totem is a plant or animal or other thing, the hereditary friend and ally—of the kin—but all plant or animal allies of individuals or of magical societies are not totems. Though the attitude of a private person to his nagual, or of a magical society to its protective animal, may often closely resemble the attitude of the group to its hereditary totem, still, the origin of this attitude of respect may be different in each case.

This is obvious, for the individual or society deliberately adopts an animal protector and friend, usually suggested in a dream, after a fast, whereas we can scarcely conceive that the totem was deliberately adopted by the first members of the first totem groups. Savages look on animals as personalities like themselves, but more powerful, gifted with more wakan, or mana, or cosmic rapport: each man, therefore, and each organised magical society, looks out for, and, for some reason of dream or divination, adopts, a special animal friend. But it is hard to believe that the members of a primeval human group of unknown antiquity, consciously and deliberately made a compact to adopt, and for ever be faithful to—this or that plant, animal, element, or the like—to be inherited in the

---

[188] So also Mr. Hartland writes, Man, 1902, No. 84. But manitu is perhaps too wide and vague a term: it usually connotes anything mystical or supernormal.

female line. For, on this plan, the group, say Wolves, instantly loses the totem it has adopted.

We cannot prove that it was not so, that a primitive group of rudimentary human beings did not make a covenant with Bear, or Wolf, as Israel did with Jehovah, and as an individual savage does with his nyarong, or nagual, or manitu. This covenant, if made and kept by each group, would be the Origin of Totemism. But, with female descent, the covenant could not be kept. I am not certain that this theory, involving joint and deliberate selection and retention of a totem, by a primeval human group, has ever been maintained, unless it be by Mr. Jevons. 'The primary object of a totem alliance between a human kin and an animal kind is to obtain a supernatural ally against supernatural foes.'[189] The term 'supernatural' seems here out of place—both the animal kind and the human kin being natural; and one has a difficulty in conceiving that very early groups of kin would make, and would adhere to, such alliances. Indeed, how could they adhere to their totems, when these descended through women of alien totem groups? But there seems to be nothing otherwise impossible or self-contradictory in this theory; nor can it be disproved, for lack of evidence. Only such theories as are self-contradictory, or inconsistent with the known and admitted facts of the case, are capable of absolute disproof.

It may, of course, be objected here that, though totems, in actual savage society, descend sometimes in the female line, still, descent in the male line may be the original rule; and that thus a group, like an individual, could seek, make a covenant with, and cleave to a grub, or frog, or lizard, as a supernatural ally. But, for reasons already indicated, in an earlier part of this work, I conceive that, originally, totems descended in the female line only. One reason for this opinion is that, as soon as descent of the totem comes by the male line, a distinct step in the upward movement towards civilisation and a settled life is made. It is not very probable that the backward step, from reckoning by male lineage to descent in the female line, has often been taken. On the other hand, tribes which now inherit the totem in the male line, exhibit in their institutions many survivals of female descent. An instance is that of the Mandans, as recorded by Mr. Dorsey.[190] Among the Melanesians, where female descent still exists, there is at work the most obvious tendency towards descent through males, as Dr. Codrington proves in an excellent work on that people. Dr.

---

[189] Introduction to the History of Religion, p. 214. Major Powell has said something to the same effect, but that was in a journal of 'popular science.'
[190] Bureau of Ethnology, 1893-1894, p. 241.

Durkheim, too, has pointed out the traces of uterine descent among the Arunta, who now reckon in the male line.[191] On the other hand, where we find descent in the male line, I am not aware that we discover signs of movement in the opposite direction. In this opinion that, as a general rule, descent was reckoned in the female, not the male line, originally, I have the support of Mr. E. B. Tylor.[192] For these reasons the hypothesis of the selection of and covenant with a 'supernatural ally,' plant or animal, by the deliberate joint action of an early group, at a given moment, involving staunch adherence to the original resolution, rather strains belief; and a suggestion perhaps more plausible will be offered later.

THE WORD 'TOTEM'

As to the precise original meaning and form of the word usually written 'totem'—whether it should be 'totam,' or 'toodaim,' or 'dodaim,' or 'ododam,' or 'ote,' philologists may dispute.[193] They may question whether the word means 'mark,' or 'family,' or 'tribe,' or clay for painting the family mark.[194] When we here use the word 'totem' we mean, at all events, the object which gives its name to a group of savage kindred, who may not marry within this hereditary name. In place of 'totem' we might use the equivalent murdu of the Dieri, or gaura of the Kunundaburi.[195]

THE TOTEM 'CULT'

The 'cult,' if it deserves to be called a 'cult,' of the totem, among savages, is not confined to abstention from marriage within the name. Each kin usually abstains from killing, eating, or in any way using its totem (except in occasional ceremonies, religious or magical), is apt to claim i descent from or kindred with it, or alternations of metamorphosis into or out of it, and sometimes uses its effigy on memorial pillars, on posts carved with a kind of genealogical tree, or tattoos or paints or scarifies it on the skin—in different cases and places.

---

[191] L'Année Sociologique, v. 93, 99, 100. As far as the proof rests on Arunta traditions, I lay no stress upon it.
[192] J. A. I. vol. xviii., no. 3, p. 254.
[193] Frazer, Totemism, p. 1.
[194] Major Powell, Man, 1901, no. 75.
[195] Howitt, J. A. I. xx. 40-41, 1891.

To what extent the blood-feud is taken up by all members of the slain man's totem, I am not fully aware: it varies in different regions. The eating or slaying of the totem, by a person of the totem name, is in places believed to be punished by disease or death, a point which the late Mr. J. J. Atkinson observed among the natives of New Caledonia (MS. penes me). Mr. Atkinson happened to be conversing with some natives on questions of anthropology, when his servant brought in a lizard which he had killed. On this one of the natives exhibited great distress, saying, 'Why have you killed my father? we were talking of my father, and he came to us' The son (his name was Jericha) then wrapped the dead lizard up in leaves, and reverently laid the body in the bush. This was not a case like that of the Zulu Idhlozi, the serpents that haunt houses, and are believed to be the vehicles of the souls of dead kinsfolk. The other natives present had for their 'father,' one, a mouse, the other a pigeon, and so on. If any one ate his animal 'father,' sores broke out on him, and Mr. Atkinson was shown a woman thus afflicted, for having eaten her 'father.' But I do not find, in his papers, that a man with a mouse for father might not marry a woman of the mouse set, nor have I elsewhere been able to ascertain what is New-Caledonian practice on this point.[196] When Mr. Atkinson made these observations (1874), he had only heard of totems in the novels of Cooper and other romancers.

'TOTEM GODS'

This example is here cited because, as far as I am aware, no other anthropologist has observed this amount of Totemism in New Caledonia. Students are divided into those who have a bias in favour of finding totemism everywhere; and those who aver, with unconcealed delight, that in this or the other place there are no totems. Such negative statements must always be received with caution. An European may live long among savages before he really knows them; and, without possessing totemism in full measure, many races retain obvious fragments of the institution.

---

[196] The Marquis d'Eguilles kindly sends me extracts from an official 'Notice sur la Nouvelle-Calédonie,' drawn up for the Paris Exhibition of 1900. The author says that the names of relationships are expressed, by the Kanaka, 'in a touching manner.' One name includes our 'uncle' and 'father,' another our 'mother' and 'aunts;' another name includes our 'brothers,' 'sisters,' and 'cousins.' This, of course, is 'the classificatory system.' About animal 'fathers' nothing is said.

Mr. Tylor has censured the use of the terms 'totems' and 'totem clans' with respect to the Fijians and Samoans, where certain animals, not to be eaten, are believed to be vehicles or shrines of certain gods. It is a very probable conjecture (so probable, I think, as almost to amount to a certainty), that the creatures which are now the shrines of Fijian or Samoan gods of the family, or of higher gods, were once totems in an earlier stage of Samoan and Fijian society and belief. As I have said elsewhere, 'in totemistic countries the totem is respected himself; in Samoa the animal is worshipful because a god abides within him. This appears to be a theory by which the reflective Samoans have explained to themselves what was once pure Totemism.'[197] But I must share in Mr. Tylor's protest against using the name of 'totem' for a plant or animal which is regarded as the shrine of a god. Such thorough totemists as some of the North American Indians, or the Australians, do not explain their totems as the shrines of gods, for they have no such gods to serve as explanations. That myth appears to be the Samoan or Fijian way of accounting for the existence of worshipful and friendly plants and animals.

Thus, at all events, and unluckily, the phrase 'the totem-god' is introduced into our speculations, and the cult of the 'totem-god' is confused with the much more limited respect paid by savages to actual totems. However attractive the theory of 'the totem-god' may be, we cannot speak of 'totems' where a god incarnate in a plant or animal is concerned. Such a deity may be a modified survival of Totemism, but a totem he is not. Moreover, it is hardly safe to say that, in the Samoan case, the god is 'developed from a totem;' we only know that the god has got into suspiciously totemistic society. On the whole, we cannot be too cautious in speaking of totems and Totemism: and we must be specially careful not to exaggerate the more or less religious respect with which totems are, in many cases, regarded. The Australians, as far as they have the idea of a creative being, Baiame, Nooreli, and so forth, do not regard their totems as shrines or incarnations of him. That appears to be the speculation of peoples who, probably by way of animism, and ancestor-worship, are already in the stage of polytheism. Totems, in their earliest known stage, have very little to do with religion, and probably, in origin, had nothing really religious about them.

---

[197] Tylor, Remarks on Totemism, pp. 141-143. Myth, Ritual, and Religion, ii. 56-58. Turner's Samoa, p. 17 (1884).

SAVAGE SPECULATIONS AS TO THE ORIGIN OF TOTEMISM

Peoples who are still in the totemistic stage, as we have seen, know nothing about the beginnings of the institution. All that they tell the civilised inquirers is no more than the myth handed down by their own tradition. Thus the Dieri or Dieyrie, in Australia, say that the totems were appointed by the ancestors, for the purpose of regulating marriages, after consultation with Mura Mura, or with 'the' Muramura. The Woeworung, according to Mr. Howitt, have a similar legend.[198] It is not necessary here to ask whether Mura Mura is 'the Supreme Being' (Gason, Howitt), or 'ancestral spirits' (Fison).[199] The most common savage myth is of the Darwinian variety, each totem kin is descended from, or evolved out of, the plant or animal type which supplies its totem. Again, as in fairy tales, a woman gave birth to animals, whence the totem kins derive their descent. In North-West America, totems are often accounted for by myths of ancestral heroes. 'The Tlingit' (Thlinket) 'hold that souls of ancestors are reborn in children, that a man will be reborn as a man, a wolf as a wolf, a raven as a raven.' Nevertheless, the totems are regarded as 'relatives and protectors,' and it is explained that, in the past, a human ancestor had an adventure with this or that animal, whence he assumed his totem armorial bearings.[200] In precisely the same way a myth, a very late myth, was invented, about the adventure of a Stewart with a lion, to account for the Lyon of the Stewarts.[201] The Haidas and Thlinkets, believing as they do that human souls are reborn human, cannot hold that a bestial soul animates a man, say, of the Raven totem.

The Arunta, on the other hand, suppose that the souls of animals which evolved into human beings, are reincarnated in each child born to the tribe. 'Two clans of Western Australia, who are named after a small opossum and a little fish, think that they are so called because they used to live chiefly on these creatures.'[202] This myth has some support in modern opinion: the kins, it is argued, received their totem names from the animals and plants which

---

[198] Howitt, On the Organisation of Australian Tribes, p. 136, note, 1889.
[199] The Mura Mura appear really to answer to the fabled ancestors of the Arunta, but are addressed in prayers. Cf. Miss Howitt, Folk Lore, January 1903.
[200] Tylor, Remarks on Totemism, p. 134.
[201] So also to explain the crest of the Hamiltons, the Skenes, and many others.
[202] Frazer, Totemism, p. 7.

mainly formed their food supply; though now their totems are seldom eaten by them. These legends, and others, are clearly ætiological myths, like the Samoan hypothesis that gods are incarnate in the totems. The myths merely try to explain the original connection between men and totems, and are constructed on the lines of savage ideas about the relations of all things in the universe, all alike being personal, and rational, and capable of interbreeding, and of shape-shifting. Certain Kalamantans of Sarawak will not eat a species of deer, because 'an ancestor became a deer of this kind.'[203] All such fables, of course, are valueless as history; and, in the savage state of the intellect, such myths were inevitable.

MODERN THEORIES

Mr. McLennan himself at first had a theory, which, as far as I heard him speak of it, was more or less akin to my own. But he abandoned it, says his brother, Mr. Daniel McLennan, for reasons that to him appeared conclusive. I ought to mention that Mr. A. H. Keane informed me, several years ago, that he had independently evolved a theory akin to mine, of which, as it then stood, I had published some hint. (For a statement of Mr. Keane's theory see our Preface.) In 1884[204] I wrote, 'People united by contiguity, and by the blind sentiment of kinship not yet brought into explicit consciousness, might mark themselves by a badge, and might thence derive a name, and, later, might invent a myth of their descent from the object which the badge represented.' But why should such people mark themselves by a badge, and why, if they did, should the mark be, not a decorative or symbolic pattern, but the representation of a plant or animal? These questions I cannot answer, and my present guess is not identical with that of 1884.

Meanwhile let us keep one point steadily before our minds. Totemism, at a first glance, seems a perfectly crazy and irrational set of beliefs, and we might think, with Dr. Johnson, that there is no use in looking for reason among the freaks of irrational people. But man is never irrational. His reason for doing this, or believing that, may seem a bad reason to us, but a reason he always had for his creeds and conduct, and he had a reason for his totem belief, a reason in congruity with his limited knowledge of facts, and with his theory of the universe. For all things he wanted an explanation. Now what he wanted a reason for, in Totemism, was the nature and

---

[203] Hose and McDougall, J. A. I. xxxi. 193, 1901.
[204] Custom and Myth, p. 262.

origin of the connection between his own and the neighbouring groups, and the plant or animal names which they bore. Messrs. Spencer and Gillen write, 'what gave rise, in the first instance, to the association of particular men with particular animals and plants, it is impossible to say.' But it is not impossible to guess, with more or less of probability. The connection once established, savages guessed at its origin: their guesses, as always, were myths, and were of every conceivable kind. The myth of descent from or kinship with the animal or plant, the Darwinian myth, does not stand alone. Every sort of myth was fashioned, was believed, and influenced conduct. Our business is to form our own guess as to the original connection between men and their totems: a guess which shall be consistent with human nature.

MR. MAX MÜLLER'S THEORY

Many such guesses by civilised philosophers exist. We need not dwell long on that of Mr. Max Müller, akin, as it is, to my own early conjecture, 'a totem is a clan mark, then a clan name, then the name of the ancestor of the clan, and lastly the name of something worshipped by the clan.'[205] We need not dwell on this, because the kind of 'clan mark' on a pillar outside of the quarters of the clan, in a village, is peculiar to North America, and to people dwelling in fixed settlements. Among the nomadic Australians, we have totemism without the settlements, without the totem pillar, without the 'clan mark,' on the pillar, which, thus, cannot be the first step in Totemism. Again, the 'clan name,' or group name, must be earlier than the 'clan mark,' which merely expresses it, just as my name is prior to my visiting card, or as the name of an inn, 'The Red Lion,' is prior to the sign representing that animal. Obviously we have to ask first, whence comes the clan name, or group name?

THE THEORY OF MR. HERBERT SPENCER

In a passage on animal-worship, Mr. Herbert Spencer (unless I misconceive him) advances a theory of the origin of Totemism. True, he does not here speak of totems, but he suggests an hypothesis to explain why certain stocks claim descent from animals, and why these animals are treated by them with more or less of religious regard. Actual men, in savagery, are often called by

---

[205] Contributions to the Science of Mythology, i. 201.

'animal nicknames,' and we cannot be surprised if the savage ... gets the idea that an ancestor named 'the tiger' was an actual tiger ... Inevitably, then, he grows up believing that his father descended from a tiger—thinking of himself as one of the tiger stock.[206]

It were superfluous to dwell on this theory. Totem names are group names; and, as they occur where group names are derived from the mother, they cannot have originated in the animal nicknames of individual dead grandfathers. The names of the dead are usually tabued and forgotten; but that is of no great moment. The point is that such group names are derived through mothers, in the first instance, not through male founders of families.[207]

No theory which starts from an individual male ancestor, and his name bequeathed to his descendants, can be correct. That Mr. Spencer's does start in this way may be inferred from the following text: 'commonly the names of the clans which are forbidden to intermarry, such as Wolf, Bear, Eagle, Whale, &c., are names given to men, implying, as I have before contended (170-173), descent from distinguished male ancestors bearing those names—descent which, notwithstanding the system of female kinship, was remembered when there was pride in the connection.'

A brief-lived joy in the name of which the male ancestor's descendants were proud, left them, in the second generation, under exogamy and female kin. Thus my father was nicknamed 'Tiger.' Proud of the title, I call myself Tiger. But I must marry a woman who is Not-Tiger, and my offspring are Not-Tigers. My honour hath departed!

MR. FRAZER'S THEORIES

The hypotheses of Mr. J. G. Frazer are purely provisional. He starts from the idea, so common in Märchen, of the person whose 'soul,' 'life,' or 'strength' is secretly hidden in an animal, plant, or other object. The owner of the soul wraps the 'soul-box' up in a mystery, it is the central secret of his existence, for he may be slain by any one who can discover and destroy his 'soul-box.' Next Mr. Frazer offers many cases of this actual belief and practice among savage and barbarous peoples; and, as a freak or survival, the idea is found even among the civilised. We meet the superstition in the Melanesian group of islands (where Totemism is all but extinct), and perhaps among the Zulus, with their serpent Idhlozi, whose life

---

[206] The Principles of Sociology, i. 362, 1876.
[207] The whole passage will be found in the work cited. Vol. i. 359-368.

is associated with their own. Mr. Atkinson's New-Caledonians, however, did not think that death inflicted on their animal 'fathers' involved danger to themselves, though it distressed them, as an outrage to sentiment. Then we have the 'bush-souls' (one soul out of four in the possession of each individual), among the natives of Calabar. These souls, Miss Kingsley wrote, are never in plants, but always in wild beasts, and are recognisable only by second-sighted men. The 'bush-soul' of a man is often that of his sons: the daughters often inherit the mother's 'bush-soul:' or children of both sexes may take the bush-soul of either father or mother. The natives will not injure their bush-soul beasts. Nothing is known as to prohibition of marriage between persons of the same bush-soul. Here we have really something akin to the totem, the bush-souls being hereditary, at least for one generation. But this is among a house-dwelling, agricultural people, far above the state of real savagery: not among a 'primitive' people.

The Zapotecs of Central America, again, choose, by a method of divination, 'a tona or second self,' an animal, for each child, at its birth. It is, by the nature of the case, not hereditable. The nagual, usually a beast, of each Indian of Guatemala is well known; and is discovered, on the monition of a dream, by each individual. Therefore it cannot be hereditable. The sexes, in Australia, have each a friendly and protecting species of animal; say a Bat for all men, a Nightjar for all women: indeed, in Australia, all the elements of nature have their place in the cosmic tribe. To injure the animal of either sex, is to injure one of the sex. There is no secret about the matter.

Mr. Frazer then argues, 'the explanation which holds good of the one' (say 'the sex totem,' or 'personal totem'), 'ought equally to hold good of the other' (the group totem). 'Therefore, the reason why a tribe' (I venture to prefer 'group,' or 'kin,' as there are many totems in each 'tribe') 'revere a particular species of animals or plants ... and call themselves after it, would seem to be a belief that the life of each individual of the tribe is bound up with some one plant or animal of the species, and that his or her death would be the consequence of killing that particular animal or destroying that particular plant.' Mr. Frazer thinks that 'this explanation squares well' with Sir George Grey's description of a Kobong or totem in Western Australia. There, a native gives his totem 'a fair show' before killing it, always affording it a chance of escape, and never killing it in its sleep. He only does not shoot his kindred animal sitting, and his plant he only spares 'in certain circumstances, and at a particular period of the year.' Mr. Frazer writes that as the man does not know which individual of the species of plant or animal 'is

114

specially dear to him, he is obliged to spare them all, for fear of injuring the dear one.' But the man, it seems from Grey's account, does not 'spare' any of them; he kills or plucks them, 'reluctantly,' and in a sportsmanlike manner, 'never without affording them a chance of escape.' In a case of Sir George Grey's, the killing of a crow hastened the death of a man of the Crow totem, who had been ailing for some days. But the Australians do not think that to kill a man's totem is to kill the man. Somebody's totem is killed whenever any animal is slain. Mr. Frazer now finds that the Battas, for example, 'do not in set terms affirm their external soul to be in their totems,' and I am not aware that any totemists do make this assertion. They freely offer all other sorts of mythical explanations as to what their totems originally were, as to the origin of their connection with their totems, but never say that their totems are their 'soul-boxes.'

Mr. Frazer has an answer to this objection. 'How close must be the concealment, how impenetrable the reserve in which he' (the savage) 'hides the inner keep and citadel of his being.' The Giant, in the Märchen, tries to keep the secret of his 'soul-box,' much more then does 'the timid and furtive savage.' 'No inducement that can be offered is likely to tempt him to imperil his soul by revealing its hiding-place to a stranger. It is, therefore, no matter for surprise that the central mystery of the savaged life should so long have remained a secret, and that we should be left to piece it together from scattered hints and fragments, and from the recollections of it which linger in fairy tales.'

On reflection, we cannot but see the flaw in this reasoning. No savage has revealed to European inquirers that his totem is his 'soul-box.' But every other savage knows his fatal secret. Every savage, well aware that his own totem is the hiding-place of his soul, knows that the totems of his enemies are the hiding-places of their souls. He wants to kill his enemies, and he has an easy mode of doing so, to shoot down every specimen of their totems. His enemies will then die, when he is lucky enough to destroy their 'soul-boxes.' Now I am not aware, in the destructive magic of savages, of a single case in which a totem is slain, or tortured for the purpose of slaying or torturing a man of that totem. All other sorts of sympathetic magic are practised, but where is the evidence for that sort, which ought to be of considerable diffusion?[208] The supposed 'secret' of savage life is no secret to other savages. Each tells any inquirer what his 'clay' or totem is. He blazons his totem proudly. The nearest approach to invidious action, against a totem,

---

[208] I am haunted by the impression that I have met examples, but where I know not.

with which I am acquainted, is the killing by the Kurnai women, of the men's 'sex totem,' when the young men are backward wooers. The purpose is to produce a fight between lads and lasses, a rude form of flirtation, after which engagements, or elopements, are apt to follow.[209]

Mr. Frazer tentatively suggests another, a rival or a subsidiary solution of the problem, to which reference has already been made. Among the Arunta and other tribes, 'the totemic system has a much wider scope, its aim being to provide the community with a supply of food and all other necessaries by means of certain magical ceremonies, the performance of which is distributed among the various totem groups.' That is to say, these totemic magical ceremonies now exist for the purpose of propagating, as part of the food supply, animals or vegetables, which, by the former theory, were the secret receptacles of the lives of the tribesmen. To kill and eat these sacred receptacles would endanger the lives of the tribesmen, but to risk that is quite in accordance with the practical turn of the Arunta mind. Mr. Frazer has, however, suggested a possible method of reconciling his earlier hypothesis—that a totem was a soul-box—with his later theory, that the primal object of totem groups was to breed their totems for food.[210]

Mr. Frazer observes, 'It is not as yet clear how far the particular theory of Totemism suggested by the Central Australian system is of general application, and ... in the uncertainty which still hangs over the origin and meaning of Totemism, it seems scarcely worth while to patch up an old theory which the next new facts may perhaps entirely demolish.' He then cites the Arunta belief that their ancestors of 'the dream time' (who were men evolved out of animals or plants, these objects being their totems) kept their souls (like the Giant of the fairy tale) in stone churingas (a kind of amulets) which they hung on poles when they went out hunting. We have thus a va-et-vient between each man, and the spirit of the plant or animal out of which he, or his human ancestor, was evolved. That spirit (in origin the spirit of an animal or plant) is now handed down with the stone churinga, and is reincarnated in each child, who is thus, an incarnation of the original totem. Such is the Arunta theory, and thus each living Arunta is the totem's soul-box, while, to savage reasoners, the totem soul may, perhaps, seem also to tenant simultaneously each plant or animal of its species.

---

[209] Howitt, Journal of the Anthropological Institute, xviii. 58.
[210] Golden Bough, iii. 416, note 3.

This is a theory of Totemism;[211] but, so far, we only know the facts on which it is based among one extraordinary tribe of anomalous development. We have still to ask, what was the original connection of the men with the plants and animals, which the Arunta explain by their myth? Was that connection originally one of magic-working, by each group, for its totem species, and, if so, why or how did the groups first select their plants and animals? Mr. Haddon's theory, presently to be criticised, may elucidate that point of departure.

SUGGESTION OF MR. N. W. THOMAS

As I am writing, a theory, or suggestion, by Mr. N. W. Thomas appears in Man (1902, No. 85). Mr. Thomas begins with the spirit which dwells in an African fetich, and becomes the servant of its owner. The magical apparatus 'may be a bag of skin containing parts of various animals. Such an animal may be the familiar of the owner, his messenger, or an evil spirit that possesses him;' similar beliefs are held about the wer-wolf. Now the American-Indian has his 'medicine bag.' 'The contents are the skin, feathers, or other part of the totem animal.'

Distinguo: they are parts, not of the 'totem animal,' but of the adopted animal of the individual, often called his manitu. If we say 'the totem animal,' we beg the question; we identify the totem with the manitu of the individual. It may be true, as Mr. Thomas says, that 'the basis of individual Totemism seems to be the same as that of fetishism,' but I am not discussing 'individual Totemism,' but real group Totemism. Mr. Thomas also is clear on this point, but, turning to Australia, he says that 'the individual totem seems to be confined to the medicine-man.' From information by Mrs. Langloh Parker, I doubt the truth of this idea. A confessedly vague reminiscence of Mr. Rusden does not help us. Speaking of an extinct tribe on the Hunter River, N.S.W., he says that he 'does not recollect all their class divisions, Yippai' (Ippai), 'and Kombo' (Kumbo). 'Apropos of the generic names' (whatever these may be) 'the Geawe-gal had a superstition that every one had within himself an affinity to the spirit of some beast, bird, or reptile. Not that he sprang from the creature in any way' (as is a common totemic myth), 'but that the spirit which was in him was akin to that of the creature.' This is

---

[211] It is possible that I have failed to understand the mode of reconciling the two hypotheses, and Mr. Frazer is not to be understood as committed to either or both in the present state of our information.

vague. Mr. Rusden does not say that his native informant said, that the 'spirit' was the man's totem in each case.[212] But Mr. Thomas, on this evidence, writes: 'This belief suggests that the interpretation suggested for individual Totemism can also be applied to clan Totemism,' apparently because, among the extinct tribe, not only sorcerers, but, in this case, every one was the receptacle of an animal (not a plant) spirit. But obviously the animal spirits of the Geawe-gal may be the spirits—not of their group totems, if they had any—but of their individual manitus, which we do not know to be confined to sorcerers. Every one is a sorcerer, better or worse, in a society where every one works magic.

Next, the wer-wolf has a way of returning 'to look at' (to eat, I think) the body of his victim. Now in North Queensland, as in Scotland, the body of a dead man is surrounded with dust or ashes (flour in Scotland), and the dust is inspected, to find the tracks of some bird or animal.[213] From such marks, if any, 'the totem of the malefactor is inferred.' The malefactor is the person who, by the usual superstition, is thought magically to have caused the death of the tribesman. 'These facts seem best interpreted if we suppose that in North Queensland the sorcerer is believed to return in animal form, and that the form is that of his totem, for in no other way does it seem possible to identify the man's totem by observing the footsteps.'

Is the man's group totem meant? If so, the process could not identify 'the malefactor,' there are hundreds of men of his totem. Is his manitu or 'individual totem' meant? Then the process might be successful, but has no concern with the origin of hereditary kin-Totemism. Indeed Mr. Thomas 'leaves the applicability of the theory to group Totemism for subsequent consideration.' We shall show—indeed, in Mr. Herbert Spencer's case we have shown—the difficulty of deriving kin-Totemism from the manitu, or 'obsessing spirit' if Mr. Thomas pleases, of the individual. This point, as is said, Mr. Thomas reserves for later consideration.

DR. WILKEN'S THEORY

We now come to a theory which exists in many shapes, but in all is vitiated, I think, by the same error of reasoning. Mr. Tylor, however, has lent at least a modified approval to the hypothesis as mooted by the late Dutch anthropologist, Dr. Wilken, of Leyden.

---

[212] Kamilaroi and Kurnai, p. 280.
[213] J. A. I. xiii. 191, note 1.

Mr. Tylor writes, 'if it does not completely solve the totem problem, at any rate it seems to mark out its main lines.' Unluckily the hypothesis of Dr. Wilken is perhaps the least probable of all. The materials are found, not in a race so comparatively early as the Australians or Adamanese, but among the settled peoples of Malay, Sumatra, and Melanesia. By them, in their Tables of Precedence, 'the Crocodile is regarded as equal in rank to the Dutch Resident.' Crocodiles are looked on as near kinsmen of men, who, when they die, expect to become crocodiles. To kill crocodiles is murder. 'So it is with tigers, whom the Sumatrans worship and call ancestors.'

Mr. Tylor observes, 'Wilken sees in this transmigration of souls the link which connects Totemism with ancestor-worship,' and thinks that Dr. Codrington's remarks on Melanesian ways add weight to this opinion. In Melanesia, as Dr. Codrington reports, an influential man, before his death, will lay a ban, or tabu, on something, say a banana, or a pig. He says that he 'will be in' a shark, a banana, a bird, a butterfly, or what not. Dr. Codrington's informant, Mr. Sleigh of Lifu, says 'that creature would be sacred to his family,' they would call it 'papa,' and 'offer it a young cocoa-nut.' 'But they did not adopt thus the name of a tribe.' The children of papa, who chose to be a butterfly (like Mr. Thomas Haynes Bailey) do not call themselves 'The Butterflies,' nor does the butterfly name mark their exogamous limit. Mr. Tylor concludes, 'an ancestor, having lineal descendants among men and sharks, or men and owls, is thus the founder of a totem family, which mere increase may convert into a totem clan, already provided with its animal name.' This conclusion is tentative, and put forth with Mr. Tylor's usual caution. But, as a matter of fact, no totem kin is actually founded thus, for example, in Melanesia. The institutions of that region, as we are to show, really illustrate the way out of, not the way into, Totemism. Moreover the theory, as expressed by Mr. Tylor in the words cited, must be deemed unfortunate because it takes for granted that 'the Patriarchal theory' of the origin of the so-called 'clan,' or totem group, is correct. A male ancestor founds a family, which swells, 'by natural increase,' into a 'clan.' The ancestor is worshipped under the name of Butterfly, his descendants, the clan founded by him, are named Butterflies. But all this can only happen where male ancestors are remembered, and are worshipped, where descent is reckoned in the male line, and where, as among ourselves, a remembered male ancestor founds a House, as Tam o' the Cowgate founded the House of Haddington. In short Dr. Wilken has slipped back into the Patriarchal theory. Now, among totemists like the Australians, ancestors are not remembered, their names are

119

tabued, they are not worshipped, they do not found families, where descent is reckoned in the female line.[214]

MISS ALICE FLETCHER'S THEORY

An interesting variant of this theory is offered, as regards the Omaha tribe of North America, by Miss Alice Fletcher, whose knowledge of the inner mind of that people is no less remarkable than her scientific caution.[215] The conclusion of Miss Fletcher's valuable essay shows, at a glance, that her hypothesis contains the same fundamental error as that of Dr. Wilken: namely, the totem of the kin is derived from the manitu, or personal friendly object of an individual, a male ancestor. This cannot, we repeat, hold good for that early stage of society which reckons descent in the female line, and in which male ancestors do not found houses, clans, names, or totem kins.

The Omaha men, at puberty, after prayer and fasting, choose manitus suggested in dreams or visions. Miss Fletcher writes, 'As totems could be obtained but in one way—through the rite of the vision, the totem of a gens must have come into existence in that manner, and must have represented the manifestation of an ancestor's vision, that of a man whose ability and opportunity served to make him the founder of a family, of a group of kindred who dwelt together, fought together, and learned the value of united strength.'[216]

This explanation obviously cannot explain the Origin of Totemism among tribes where descent is reckoned in the female line, and where no man becomes 'the founder of a family.' The Omaha, a house-dwelling, agricultural tribe, with descent in the male line, with priests, and departmental gods, a tribe, too, among whom the manitu is not hereditable, can give us no line as to the origin of Totemism. Miss Fletcher's theory demands the hereditable character of the individual manitu, and yet it is never inherited.

MR. HILL TOUT'S THEORY

Mr. Hill Tout has evolved a theory out of the customs of the aborigines of British Columbia, among whom 'the clan totems are a

---

[214] Tylor, Remarks on Totemism, pp. 146-147, 1898.
[215] The Import of the Totem, by Alice C. Fletcher, Salem Press, Mass., 1897.
[216] Op. cit. p. 12.

development of the personal or individual totem or tutelar spirit.' The Salish tribes, in fact, seek for 'sulia, or tutelar spirits,' and these 'gave rise to the personal totem,' answering to manitu, nyarong, nagual, and so forth. 'From the personal and family crest is but a step to the clan crest.' Unluckily, with descent in the female line, the step cannot be taken. Mr. Hill Tout takes a village-inhabiting tribe, a tribe of village communities, as one in which Totemism is only nascent. 'The village community apparently formed the original unit of organisation.' But the Australians, who have not come within measurable distance of the village community, have already the organisation of the totem kin. Interesting as is Mr. Hill Tout's account of the Salish Indians, we need not dwell longer on an hypothesis which makes village communities prior to the evolution of Totemism. What he means by saying that 'the gens has developed into the clan,' I am unable to conjecture. The school of Major Powell use 'gens' of a totem kin with male, 'clan' of a totem kin with female descent. Mr. Hill Tout cannot mean that male descent is being converted into descent in the female line? As he writes of 'a four-clan system, each clan being made up of groups of gentes,' he may take a 'clan' to signify what is usually called a 'phratry.'[217]

MESSRS. HOSE AND MCDOUGALL

Among other efforts to show how the hereditary totem of a group might be derived from the special animal or plant friend of an individual male, may be noticed that of Messrs. Hose and McDougall.[218] The Ibans, or Sea Dyaks of Sarawak, are probably of Malay stock, and are 'a very imitative people,' of mixed, inconsistent, and extravagant beliefs. They have a god of agriculture, and, of course, are therefore remote from the primitive; being rice-farmers. They respect nyarongs, or 'spirit helpers,' though Mr. Hose lived among them for fourteen years without knowing what a nyarong is. 'It seems usually to be the spirit of some ancestor, or dead relative, but not always so....' The spirit first appears to an Iban in a dream, in human form, and the Iban, on awaking, looks for the nyarong in any casual beast, or quartz crystal, or queer root or creeper. So far the nyarong is a fetish. Only about two per cent, of men have nyarongs. If the thing be an animal, the

---

[217] 'The Origin of the Totemism of the Inhabitants of British Columbia,' Transactions of Royal Society of Canada, second series, vol. vii., 1901-1902. Quaritch, London.
[218] J. A. I. xxxi. 196, et seq.

Iban respects the other creatures of the species. 'In some cases the cult of a nyarong will spread through a whole family or household.' Australian individuals have also their secret animal friends, like nyarongs and naguals, but these are never hereditary. What is hereditary is the totem of the group, which may not be altered, or so seldom that it would be hard to find a modern example: though changes of totems may have occurred when, in the pristine 'treks' of the race, they reached regions of new fauna and flora.

'The children and grandchildren,' our authors go on, 'among the Ibans, will usually respect the species of animals to which a man's nyarong belongs, and perhaps sacrifice fowls or pigs to it occasionally.' Of course 'primitive' man has no domesticated animals, and does not sacrifice anything to anybody. 'If the great-grandchildren of a man behave well to his nyarong, it will often befriend them just as much as its original protégé.' It is not readily conceivable that, among very early men, and where the names of the dead are tabued, the wisest great-grandchild knows who his great-grandfather was. Still, though the great-grandfather was forgotten, his nyarong—it may be said—would be held in perpetual memory, and become the totem of a group. But this is not easily to be conceded, because there would be the competition of the nyarongs of each generation to crush the ancient nyarong; moreover the totem, in truly primitive times, is not inherited from fathers, but from mothers.

Our authors say that, in some cases, 'all the members of a man's family, and all his descendants, and, if he be a chief, all the members of the community over which he rules,' may come to share in the benefits of his nyarong, and in its rites. But all this of chiefs, and great-grandchildren of a known great-grandfather, all this occurring to-day among an imitative and agricultural people, with departmental deities, and domesticated animals, cannot give us a line to the origin of Totemism among houseless nomads, who tabu the memory of their dead, and, as a rule, probably reckoned descent on the female side, so that a man could not inherit his father's totem. We must try to see how really early men became totemic. Mr. Frazer observes, 'It is quite possible that, as some good authorities incline to believe, the clan totem has been developed out of the individual totem by inheritance,' and Miss Alice C. Fletcher we have cited as holding this process to be probable in North America.[219] All such theories are based on the beliefs and customs of modern savages advancing, like the American Indians of to-day, towards

---

[219] Golden Bough, iii. 419, note 5.

what is technically styled 'barbarism.' It was not in the state of barbarism, but in a savagery no longer extant, that totemism was evolved. Totemism derived from inheritance of a male ancestor's special 'spirit-helper' is checked by the essential conditions of people who are settled, agricultural, and given to reckoning descent in the male line. No more can be produced, in such a state, than 'abortive beginnings of Totemism.'[220] Exogamy is never reached on these lines, and Totemism is behind, not in front of, all such peoples. Totemism arose in the period of the group, not of the family-founding male ancestor.

Messrs. Hose and McDougall, it is to be noted, do not say that Totemism is now being developed, in Sarawak, out of nyarongs. They only say that it, perhaps, might be so developed 'in the absence of unfavourable conditions.' If there existed 'prosperous families,' each with a nyarong, other families would dream of nyarongs, and it would become rather disreputable to have none. 'So a system of clan totems would be established.' But male kinship, agriculture, metal-working, chiefship, and large houses were certainly non-existent when Totemism was first evolved. We must not look, in such advanced society, for the origin of Totemism. In Sarawak is a houseless nomadic race, the Punans. Among them Totemism has not yet been observed, but they are so little known, that the present negative evidence cannot be regarded as conclusive. Mr. Hose knew the Ibans for fourteen years without learning what a nyarong is, and it was by, mere accident that Mr. Atkinson discovered the animal 'fathers' of the Kanakas.

MR. HADDON'S THEORY

Mr. Haddon has suggested a theory which was printed in the Proceedings of the British Association (1902). On this scheme, at a very early period, groups, by reason of their local environment, would have special varieties of food. Thus, at present, in New Caledonia, the Sea branch of a tribe has cocoa-nuts, fish of all sorts, and so forth, while the Bush branch has bananas, and other commodities, and the Sea and Bush moieties of the tribe meet at markets for purposes of barter. But, in a really primitive state, there will be no cultivation, as there is in New Caledonia. Still, a coast savage might barter crabs for a kangaroo, and, if landed property is acknowledged, owners of plum-trees, or of a spot rich in edible

---

[220] Hose and McDougall, op. cit. p. 211.

grass-seeds, might trade these away for lobsters and sea-perch.[221] Not having any idea of real cultivation, or of pisciculture (though they may and do have 'close' seasons, under tabu), the savages may set about working magic for their specialities in food. Thus it is conceivable that the fishers might come to be named 'crab-men,' 'lobster-men,' 'cuttlefish-men,' by their neighbours, whom they would speak of as 'grass-men,' 'plum-men,' 'kangaroo-men,' and so on. When once these names were accepted (I presume), and were old, and now of unknown or rather forgotten origin, all manner of myths to account for the connection between the groups and their plant and animal names would arise. When the myth declared that the plants and animals were akin to their name-giving creatures, superstitious practices would follow. We have seen two cases in which Australian totem groups averred that they were named totemically after a small species of opossum, and a fish which their ancestors habitually ate. But that is an explanatory myth. Man cannot live on opossums alone, still less on sardines.

My own guess admits the possibility of this cause of giving plant and animal names to groups, among other causes. But I doubt if this was a common cause. In Australia, everything that can be eaten is eaten by all the natives of a given area, each kindred having only a tendency to spare its own totem, while certain other tabus on foods exist. In this condition of affairs, very few groups could have a notable special variety of food, except in the case of certain fruits, grass-seeds, and insects. For these articles the season is almost as brief as the season of the mayfly or the grannom. 'When fruits is in, cats is out,' as the pieman said to the young lady. During the rest of the year, all the groups in a large area will be living on the same large variety of reptiles, roots, animals such as rats and lizards, birds and so forth. It does not seem probable that, except as between Sea and Bush parts of a tribe, there could be much specialisation in matters of diet, during the greater part of the year. Therefore, I do not think that the derivation of totem names from special articles of food can ever have been common. But local knowledge is necessary on this point. Are the totem groups of Australia settled on lands peculiarly notable for the plants and

---

[221] Mr. Haddon's theory involves the existence of barter between groups that had special articles of food. Under 'Hypothetical early groups' I show proof of the extreme hostility of adjacent groups in some regions. The merchant, with his articles of barter, would there himself be eaten. Mr. Atkinson's cook was eaten by his neighbours. Mr. Haddon does not hold that the primitive human groups were thus mutually hostile. Here we differ in opinion.

animals whose names they bear? If so, that circumstance may account for the totem names of each group, and—granting that the origin of the names is long ago forgotten, and that native speculation has explained the names by myths—the rest is easy.

It will appear, when we come to my conjecture, that it varies from Mr. Haddon's only on one point. We both begin with plant and animal names given to the various groups, from without. We then suppose (or, at least, I suppose) the origin of the names to be forgotten, and a connection to be established between the groups and their name-giving objects, a connection which is explained by myths, while belief in these gives rise to corresponding behaviour: respect for the totem, and for his human kinsfolk. The only difference is that my theory suggests several sources of the names: while Mr. Haddon offers only one source, special articles of food and barter. Kindreds, to be sure, are now named, not from what they eat (scores of things), but from the one thing which (as a kindred) they do not eat. But this, when once the myths of kinship with the totem arose, might be a later development, arising out of the myth. In essentials, my conjecture appears to be in harmony with Mr. Haddon's—the two, of course, were independently evolved.

On one point I perceive no difficulty, and no difference. It has been suggested that Mr. Haddon 'commences with the commencement,' whereas, in the hypothetical early age which we both contemplate, people had scarcely a sufficient command of language to invent nicknames. Why more command of language is needed for the application of nicknames than of names, I do not perceive. In Mr. Haddon's theory, as in mine, names already existed, names of plants and animals. In both of our hypotheses those names were transferred to human groups; in my conjecture for a variety of reasons, in his, solely from connection with special articles of food, eaten and bartered, by each group. I am not convinced that, so early, the relations between groups would admit of frequent barter: nor, as has been said, am I certain that many groups could have a very special article of food, in an age prior to cultivation. But, granting all that to Mr. Haddon, no more command of language is needed by my theory than by his. Each conjecture postulates the existence of names of plants and animals, and the transference of the names to human groups. If gesture language was prior to spoken language, in each case gesture names could be employed, as, in North America, totem names are to this day expressed in gesture language. In my own opinion, man was as human as he is to-day, when totem names arose, and as articulate. But, if he was not, gesture-language would suffice.

I shall illustrate my theory from folk-lore practice. We might

125

do the same for Mr. Haddon's. We talk of 'the Muffin man,' the man who sells muffins. We style one person 'The English Opium-Eater,' another 'The Oyster-Eater,' another 'The Irish Whiskey Drinker.' Here are the nicknames derived from the dealing in, or special consumption of, articles of food.

Many others occur in my folk-lore and savage lists of nicknames. They all imply at least as much command of language, as the names, ultimately totem names, given, for various reasons, in my theory. Thus Mr. Haddon and myself do not seem to me to differ on this point: his theory goes no further back in culture than mine does: nay, he assumes that barter was a regular institution, which implies a state of peace, almost a state of co-operation.

AN OBJECTION TO ALL THE THEORIES ENUMERATED

Not one of the theories here summarised, except the Dieri and Woeworung myth, explains why members of the various totem kins are exogamous, may not marry other members of the name. Suppose you do get your totem name from that of a distinguished male ancestor, why may you not marry another descendant? If because the common name, say Emu, is taken to indicate some sort of blood-relationship, why may you not marry a blood relation, even if there be no traceable kinship between you and her? A Douglas may marry a Douglas, a Smith may marry a Smith; but an Emu is often capitally punished if he marries an Emu. Suppose you get your totem name from the beast for which you do magic. Why may you not marry a person who bears the name of the same beast, and whose male kindred do magic for it? Because it is sacrosanct to you and her? But you are actually breeding it for the food-market. The answer must be that you may not marry a person who bears your own totem name, and is in the same branch of the Co-operative Magical Stores, because her beast and yours are in the same phratry, and phratry mates may not intermarry. But why may they not marry? The reply will probably be, because the legislator divided the previously undivided commune into two intermarrying exogamous phratries. But that theory we have shown to be untenable. Thus not one of the extant hypotheses of the origin of Totemism explains why totem kins are exogamous, unless Mr. Haddon supposes that the totem names, once given from without, came to be explained by myths asserting the sacred character and tribal kindred of the totem. Mr. Haddon has not said anything about a previous exogamous tendency in each of the groups which, by his

126

scheme, received totem names from without. By my hypothesis, these groups had already a strongly exogamous tendency, which later was hall-marked and sanctioned by the totem, with its myths and tabus. This advantage of explaining the exogamous attribute of the totem, my scheme possesses, and its rivals lack.

STATEMENT OF THE PROBLEM

Let us concentrate, now, our attention on the character of the genuine totem, the totem of the group or kin. It is not adopted by the savages on a dream-warning; each man or woman for himself or herself: nor is it chosen for each child at birth, nor by a diviner, like the nagual, bush-soul, nyarong of Sarawak, or the secret animal friend of each individual Australian. A savage inherits his group totem name. The name of any plant or animal which he may adopt for himself, or have assigned to him as a personal name, by his parents, or, so to speak, god-parents, is not his totem. My meaning is, I repeat, that my conjecture is only concerned with hereditary kin-totems and hereditary totem names of kindreds. No others enter into my conjecture as to origin. What some call 'personal totems,' adopted by the individual, or selected by others for him after his birth, such as the Calabar 'bush-soul,' the Sarawak nyarong, the Central American nagual, the Banks Island tamaniu, and the analogous special animal of the Australian tribesman (observed chiefly, as far as I know, by Mr. Howitt[222] and Mrs. Langloh Parker), do not here concern me. They are not hereditary group names.

THE AUTHOR'S OWN CONJECTURE

I now approach my own conjecture as to the origins of the genuine, hereditary, exogamous Totemism of groups of kin, real or imagined. Totemism as we know it, especially in some tribes of North America and in Australia, has certainly, as a necessary condition, that state of mind in which man regards all the things in the world as very much on a level in personality; the beasts being even more powerful than himself. Were it not so, the totem myths about human descent from beasts and plants: about friendly beasts, beasts who may marry men, and about metamorphoses, could not have been invented and believed, even to whatever extent myths are

---

[222] J. A. I. xiii.; Folk Lore, 10, 491.

believed. We may say that such beliefs are real, where they regulate conduct. So far, there is probably no difference of opinion, among anthropologists.

THE CONNECTION BETWEEN GROUPS AND TOTEMS

In all theories, the real problem is, how did the early groups get their totem names? The names, once accepted and stereotyped, implied a connection between each kindred, and the animal, plant, or other thing in nature whose name the kindred bore. Round the mystery of this connection the savage mind would play freely, and would invent the explanatory myths of descent from, and kinship with, or other friendly relations with, the name-giving objects. A measure of respect for the objects would be established: they might not be killed or eaten, except under necessity: magic might be worked by human Emus, Kangaroos, Plum-trees, and Grubs for their propagation, as among the Arunta and other tribes; or against them, to bar their ravaging of the crops, as among the Sioux. As a man should not spear a real Emu, if the Emu was his totem, so he does not, for reasons to be adduced, marry or have an amour with a woman who is also of the Emu blood. That is part of the tabu, resulting from the circumstances presently to be explained.

All these things, given the savage stage of thought, would inevitably follow from the recognised but mysterious connection between men and the plants and animals from which they were named. All such connections, to the savage, are blood-relationships, and such relationship involves the duties which are recognised and performed. But how did the early groups come to be named after the plants and animals; the name suggesting the idea of connection, and the idea of connection involving the duties of the totemist to his totem, and of the totem to the totemist?

NO 'DISEASE OF LANGUAGE'

The names, I repeat, requiring and receiving mythical explanations, and the explanations necessarily suggesting conduct to match, are the causes of Totemism. This theory is not a form of the philological doctrine, nomina numina. This is no case of disease of language, in Mr. Max Müller's sense of the words. A man is called a Cat, all of his kin are Cats. The language is not diseased, but the man has to invent some reason for the name common to his kin. It is not even a case of Folk Etymology, as when a myth is invented to

128

explain the meaning of the name of a place, or person, or thing. Thus the Loch of Duddingston, near Edinburgh, is explained by the myth that Queen Mary, as a child, used to play at 'dudding' (or skipping) stones across the water: 'making ducks and drakes.' Or again, marmalade is derived from Marie malade. Queen Mary, as a child, was seasick in crossing to France, and asked for confiture of oranges; hence Marie malade—'marmalade.' In both cases, the name to be explained is perverted. There is no real 'stone' in Duddingston—'Duddings' town,' the ton or tun of the Duddings; while 'marmalade' is a late form of 'marmalet,' a word older than Queen Mary's day.

An example of a folk etymology bordering on Totemism is the supposed descent of Clan Chattan, and of the House of Sutherland, from the Wild Cat of their heraldic crests. Now Clan Chattan is named, not from the cat, but from Gilla Catain, 'the servant of Saint Catan,' a common sort of Celtic personal name, as in Gilchrist.[223] The Sutherland cat-crest is, apparently, derived from Catness, or Caithness. That name, again, is mythically derived from Cat, one of the Seven Sons of Cruithne who gave their names to the seven Pictish provinces, as Fib to Fife, and so on. These Seven Sons of Cruithne, like Ion and Dorus in Greece (Ionians, Dorians), are mere mythical 'eponymoi' or name-giving-heroes, invented to explain the names of certain districts. In Totemism this is not so. Not fancied names, like Duddingstone, or Marmalade, are, in Totemism, explained by popular etymologies. Emu, Kangaroo, Wolf, Bear, Raven, are real, not perverted names, the question is, why are these names borne by groups of human beings? Answers are: given in all the numerous savage myths, whether of a divine ordinance (Dieri, Woeworung) or of descent and kinship, of intermarriage with beasts, or of adventures with beasts, or of a woman giving birth to beasts, or of evolution out of bestial types, and all these myths suggest mutual duties between men and their totems, as between men and their human kinsfolk. It will be seen that here no disease of language is involved, not even a volks-etymologie (a vera causa of myth).

If it could be shown by philologists that many totem names originally meant something other than they now do, and that they were misunderstood, and supposed to be names of plants and animals, then 'disease of language' would be present. Thus λύκος and ἄρκτος have really been regarded, as meaning, each of them 'the bright one,' and the Wolf Hero of Athens, and the Bear of the Arcadians, have been explained away, as results of 'disease of

---

[223] Macbain, Etymological Dictionary, 1896, quoting manuscript of 1456.

language.' But nobody will apply that obsolete theory to the vast menagerie of savage totem names.

HYPOTHETICAL EARLY GROUPS BEFORE TOTEMISM

But, discarding this old philological hypothesis, how did the pristine groups get their totem names? We ought first to return to our conjecture as to what these pristine groups were like. They must have varied in various environments. Where the sea, or a large lake, yields an abundant food-supply, men are likely to have assembled in considerable numbers, as 'kitchen middens' show, at favourable stations. In great woods and jungles the conditions of food-supply are not the same as in wide steppes and prairies, especially in the uniform and arid plateaus of Central Australia. Rivers, like seas and lakes, are favourable to settlement; steppes make nomadism inevitable, before the rise of agriculture. But, if the earliest groups were mutually hostile, strongly resenting any encroachment on their region of food-supply, the groups would necessarily be small, as in Mr. Darwin's theory of small pristine groups, the male, with his females, daughters, and male sons not adult.[224] A bay, or inlet, or a good set of pools and streams, would be appropriated and watchfully guarded by a group, just as every area of Central Australia has its recognised native owners, who wander about it, feeding on grubs, lizards, snakes, rats, frogs, grass-seeds, roots, emus, kangaroos, and opossums.

The pristine groups, we may be allowed to conjecture, were small. If they were not, the hypotheses which I venture to present are of no value, while that of Mr. Atkinson shares their doom. Mr. McLennan, as far as one can conjecture from the fragments of his speculations, regarded the earliest groups as at least so large, and so bereft of women, that polyandry was the general rule. Mr. Darwin, on the other hand, began with Polygyny and Monogamy, 'jealousy determining the first stage.'[225] This meant that there was a jealous old sire, who kept the women to himself, as in Mr. Atkinson's theory. As we can scarcely expect to reach certainty on this essential point, anthropology becomes (like history in the opinion of a character in Silas Marner) 'a process of ingenious guessing.' But,

---

[224] Descent of Man, ii. 362.
[225] Studies in Ancient History, second series, p. 50.

embarking on conjecture, I venture to suggest that the problem of the commissariat must have kept the pristine groups very small.[226]

They 'lived on the country,' and the country was untilled. They subsisted on the natural supplies, and the more backward their material culture, the sooner would they eat the country bare, as far as its resources were within their means of attainment. One can hardly conceive that such human beings herded in large hordes, rather they would wander in small 'family' groups. These would be mutually hostile, or at least jealous: they could scarcely yet have established a modus vivendi, and coalesced into the friendly aggregate of a local tribe, such as Arunta, Dieri, Urabunna, and so on. Such tribes have now their common councils and mysteries lasting for months among the Arunta. We cannot predicate such friendly union of groups in a tribe, for the small and jealous knots of really early men; watchfully resenting intrusion on their favourite bays, pools, and hunting of browsing grounds. As to marriage relations, it is not improbable that 'sexual solidarity' (as Mr. Crawley calls it), the separation of the sexes—the little boys accompanying the men, the little girls accompanying the women— and perhaps that 'sexual tabu,' coupled with the jealousy of male heads of groups, may already have led to prohibition of marriage within the group, and to raids for women upon hostile groups. The smaller the group, the more easily would sexual jealousy prohibit the lads from dealings with the lasses of their own group. There might thus, in different degrees, arise a tendency towards exogamy, and specially against son and mother, or father's mates, and brother and sister marriage. The thing would not yet be a sin, forbidden by a superstition, but still, the tendency might (as we have already said) run strong against marriage within each little group.

HOW THE GROUPS GOT NAMES

Up to this point we may conceive that the groups were anonymous. Each group would probably speak of itself as 'the Men' (according to a well-recognised custom among the tribes of to-day; for instance, the Gournditch-mara of Australia, mara meaning 'men'; Kurnai and Narinyeri, also mean 'the men'), while it would know neighbouring groups as 'the others,' or 'the wild blacks.' But this arrangement manifestly lacks distinctness. Even 'the others down there' is too like the vague manner in which the Mulligan

---

[226] This is the opinion not only of Mr. Darwin but of Major Powell and Mr. McGee.

indicated his place of residence. Each group will need a special name for each of its unfriendly neighbours.

These names, as likely as not, or more likely than not, will be animal or plant names, given for various reasons, perhaps, among others from fancied resemblances. It may be objected that an individual may bear a resemblance to this or that animal, but that a group cannot. But it is a peculiarity of human nature, to think that strangers (of another school eleven, say) are all very like each other, and if one of them reminds us of an Emu or a Kangaroo, all of them will. Moreover the name may be based on some real or fancied group trait of character, good or bad, which also marks this or that type of animal, such as cunning, cruelty, cowardice, strength, and so forth, and animal names may even be laudatory. We have also to reckon with the kinds of animals, plants, trees, useful flints, and other objects which may be more prevalent in the area occupied by each group; and with specialities in the food of each group's area, as in Mr. Haddon's theory. Thus there are plenty of reasons for the giving of plant and animal names, which, I suggest, were imposed on each group from without.

It is true that local names would serve the turn, if they were in use. But the 'hill-men,' 'the river-men,' 'the bush-men,' 'the men of the thorn country,' 'the rock-men,' are at once too scanty and too general. Many groups might fall collectively under each such local name. Again, it is as society moves away from Totemism, towards male kinship, and settled abodes, that local names are given to human groups, as in Melanesia, or even to individuals, as in the case of the Arunta, and the Gournditcha Mara. Among the Arunta a child is 'of' the place where he or she was born, like our de and von.[227] The piquancy of plant and animal names for groups probably hostile must also be considered. We are dealing with a stage of society far behind that of Mincopies, or Punans of Borneo, or Australians, and in imagining that the groups were, as a rule, hostile, we may or may not be making a false assumption. We are presuming that the jealousy of the elder males drove the younger males out of the group, or at least compelled them to bring in females from other groups, which would mean war. We are also assuming jealousy of all encroachments on feeding grounds. These are the premises, which cannot be demonstrated, but only put in for the sake of argument. In any case no more hostility than our and the French villages have for each other is enough to provide the giving of animal sobriquets.

As to hostility, Mr. Atkinson, in New Caledonia, had a set of labourers brought in from a distant island. Among them was a

---

[227] Spencer and Gillen, p. 57, note.

young boy, who, being employed as cook, had a good deal of popularity with his mates. He went home for a holiday, with a few men from his own island. He was put down at their little harbour, only a few miles from that of his tribe, and was instantly killed and eaten.

In 'Notice sur la Nouvelle-Calédonie' (1900) this ferocious hostility between near neighbouring groups is corroborated. It is certain death for the crew of a canoe to be driven into a harbour, however near their own, which is not their own. This is among the islanders not under the French. Count von Pfeil remarks on the violent hostility between Kanakas and others near adjacent.[228]

On this point of unfriendly sobriquets I may quote MM. Gaidoz and Sébillot.[229]

'In all ages men love to speak ill of their neighbour: to blazon him, in the old phrase of a time when our speech was less prudish, and more gay. Pleasantries are exchanged not only between man and man, but between village and village. Sometimes in one expressive word, the defect, or the quality (usually the defect), the dominant and apparently hereditary trait of the people of a race or a province is stated ... in a kind of verbal caricature.... Les hommes se sont donc blazonnés de tout le temps?

De tout le temps! MM. Gaidoz and Sébillot were not thinking of the origin of totem names, but their theory applies 'to all ages,' even the most primitive. Among French village sobriquets I note, at a hasty glance,

Largitzen	Cows	Houmeau	Frogs
Angoulême	Lizards	Artois	Dogs
Aire	Pigeons	Avalon	Birds

and villages named as eaters of:
Old Ewes
Onions
Crows.

We shall see that many Sioux groups, many English villages are blazoned, as in Mr. Haddon's theory, by the names of the things which they eat: or are accused of eating.

Thus, among very early men, the names by which the groups knew their neighbours would be names given from without. To call

---

[228] J. A. I. May 1897, p. 181.
[229] Blason Populaire de la France, p. 5. Paris, Cerf, 1884.

them 'nicknames,' is to invite the objection that nicknames are essentially derisive, and that groups so low could not yet use the language of derision. I see no reason why early articulate-speaking men (or even men whose language is gesture language) should be so modern as to lack all sense of humour, all delight in derision. But the names need not have been derisive. If these people had the present savage belief in the wakan, or mystic power of animals, the names may even have been laudatory. I ask for no more than names conferred from without, call them nicknames, sobriquets, or what you like.

We are acquainted with no race that is just entered on Totemism, unless we agree with Mr. Hill Tout that Totemism is nascent among the Salish tribe, who live in village communities. Consequently we cannot prove that early hostile groups would name each other after plants and animals. I am only able to demonstrate that, alike in English and French folk-lore, and among American tribes who reckon by the male line, who are agricultural and settled, the villages or groups are named, from without, after plants and animals, and after what they are supposed to be specially apt to use as articles of food, and also by nick-names—often derisive. What I present is, not proof that the primal groups named each other after plants and animals, but proof that among our rustics, by congruity of fancy, such names are given, with other names exactly analogous to those now used among settled savages moving away from Totemism.

ILLUSTRATION FROM FOLK-LORE

I select illustrative examples from the blason populaire of modern folk-lore. Here we find the use of plant and animal names for neighbouring groups, villages, or parishes. Thus two informants in a rural district of Cornwall, living at a village which I shall call Loughton, found that, when they walked through the neighbouring village, Hillborough, the little boys 'called cuckoo at the sight of us.' They learned that the cuckoo was the badge, in folk-lore, of their village. An ancient carved and gilded dove in the Loughton church 'was firmly believed by many of the inhabitants to be a representation of the Loughton Cuckoo,' and all Loughton folk were Cuckoos. 'It seems as if the inhabitants do not care to talk about these things, for some reason or other.' A travelled Loughtonian 'believes the animal names and symbols to be very ancient, and that each village has its symbol.' My informants think that 'some modern badges have been substituted for more ancient ones,' such as tiger and monkey. There is apparently no veneration of the local beast,

134

bird, or insect, which seems often, on the other hand, to have been imposed from without as a token of derision. Australians make a great totem of the Witchetty Grub (as Spencer and Gillen report), but the village of Oakditch is not proud of its potato grub, the natives themselves being styled 'tater grubs.' I append a list of villages (with false names[230]) and of their badges:

Hillborough	Mice	Brailing	Peesweeps
Loughton	Cuckoos	Wickley	Tigers
Miltown Mules	Fenton Rooks		

(it used to be rats)   Linton Men

Ashley	Monkeys	Oakditch	Potato grubs
Yarby	Geese	St. Aldate's	Fools
Watworth	Bulldogs		

At Loughton, when the Hillborough boys pass through on a holiday excursion, the Loughton boys hang out dead mice, the Hillborough badge, in derision. The boys have even their 'personal totem,' and a lad who wishes for a companion in nocturnal adventure will utter the cry of his peculiar beast or bird, and a friend will answer with his. If boys remained always boys (that is, savages), and if civilisation were consequently wiped out, myths about these group names of villages would be developed, and Totemism would flourish again. Later I give other instances of village names answering to totem names, and in an Appendix I give analogous cases collected by Miss Burne in Shropshire, and others, we saw, are to be found in the blason populaire of France.

It appears to me that totem group names may, originally, have been imposed from without, just as the Eskimo are really Inuits; 'Eskimo,' 'Eaters of raw flesh,' being the derisive name conferred by their Indian neighbours. Of course I do not mean that the group names would always, or perhaps often, have been, in origin, derisive nicknames. Many reasons, as has been said, might prompt the name-giving. But each such group would, I suggest, evolve animal and vegetable nicknames for each neighbouring group. Finally some names would 'stick,' would be stereotyped, and each group would come to answer to its nickname, just as 'Pussy Moncrieff,' or 'Bulldog Irving,' or 'Piggy Frazer,' or 'Cow Maitland,' does at school.

---

[230] Pseudonyms were given to avoid arousing local attention, when I put forth these facts in The Athenæum. For reasons, I retain the pseudonyms; but for the real village names see p. 173, note 1.

HOW THE NAMES BECAME KNOWN

Here the questions arise, how would each group come to know by what name each of its neighbours called it, and how would hostile groups Come to have the same nicknames for each other? Well, they would know the nicknames through taunts exchanged in battle.

> 'Run, you deer, run!'
> 'Off with you, you hares!'
> 'Skuttle, you skunks!'

They would readily recognise the appropriateness of the names, if derived from the plants, trees, or animals most abundant in their area, and most important to their food supply: for, at this hypothetical stage, and before myths had crystallised round the names, they would have no scruples about eating their name-giving plants, fruits, fishes, birds, and animals. They would also hear their names from war captives at the torture stake, or on the road to the oven, or the butcher. But the chief way in which the new group names spread would be through captured women; for, though there might as yet be only a tendency towards exogamy, still girls of alien groups would be captured as mates. 'We call you the Skunks,' or whatever it might be, such a bride might remark, and so knowledge of the new group names would be diffused. These names would adhere to groups, on my hypothesis, already exogamous in tendency, and, when the totem myth arose, the exogamy would be sanctioned by the totem tabu.[231]

TOTEMIC AND OTHER GROUP NAMES—ENGLISH AND NORTH AMERICAN INDIAN

It may seem almost flippant to suggest that this old mystery of Totemism arises only from group names given from without, some of them, perhaps, derisive. But I am able to demonstrate that, in North America, the names of what some American authorities call gentes (meaning old totem groups, which now reckon descent through the male, not the female line), actually are nicknames—in certain cases derisive. Moreover, I am able to prove that, when the names of these American gentes are not merely totem names, they

---

[231] Some objections are noticed later.

answer, with literal precision, to our folk-lore village sobriquets, even when these are not names of plants or animals. The late Rev. James Owen Dorsey left, at his death, a paper on The Siouan Sociology.[232] Among the gentes (old totem kindreds with male descent) he noted, the gentes of a tribe, 'The Mysterious Lake Tribe.' There were, in 1880, seven gentes. Three names were derived from localities. One name meant 'Breakers of (exogamous) Law.' One was 'Not encumbered with much baggage.' One was Rogues ('Bad Nation'). These three last names are derisive nicknames. The seventh name was 'Eats no Geese,' obviously a totemic survival. Of the Wahpeton tribe all the seven gentes derived their names from localities. Of the Sisseton tribe, the twelve names of gentes were either nicknames (one, 'a name of derision'), or derived from localities.

Of the Yankton gentes, five names out of seven were nicknames, mostly derisive, the sixth was 'Bad Nation' ('Rogues'), the seventh was a totem name, 'Wild Cat.' Of the Hunpatina (seven gentes), three names were totemic (Drifting Goose, Dogs, Eat no Buffalo Cows); the others were nicknames, such as 'Eat the Scrapings of Hides.' Of the Sitcanxu, there were thirteen gentes. Six or seven of their titles were nicknames, three were totemic, the others were dubious, such as 'Smellers of Fish.' The Itaziptec had seven gentes; of their names all were nicknames, including 'Eat dried venison from the hind quarter.' Of the Minikooju, there were nine gentes. Eight names were nicknames, including 'Dung Eaters.' One seems totemic, 'Eat no Dogs.' Of five Asineboin gentes the names were nicknames from the habits or localities of the communities. One was 'Girl's Band,' that is, 'Girls.'

Now compare parish sobriquets in Western England.[233] In this list of parish or village nicknames, twenty-one are derived from plants and animals, like most totemic names. We also find 'Dog Eaters,' 'Bread Eaters,' 'Burd Eaters,' 'Whitpot Eaters,' and, answering to 'Girl's Band' (Gens des Filles), 'Pretty Maidens:' answering to 'Bad Nation,' 'Rogues': answering to 'Eaters of Hide Scrapings' 'Bone Pickers': while there are, as among the Siouans, names derived from various practices attributed to the English villagers, as to the Red Indian gentes.

No closer parallel between our rural folk-lore sobriquets of village groups, given from without, and the names given from without of old savage totem groups (now reckoning in the male line,

---

[232] Report of American Bureau of Ethnology, 1893-1894, p. 213 et seq.
[233] Thirteenth Report of the Committee of Devonshire Folk-Lore, Devonshire Association for the Advancement of Science, 1895, xxvii. 61-74.

and, therefore, now settled together in given localities) could be invented. (For other examples see Appendix A.) I conceive, therefore, that my suggestion—the totem names of pristine groups were originally given from without, and were accepted (as in the case of the nicknames of Siouan gentes, now accepted by them)— may be reckoned no strain on our sense of probability. It is demonstrated that the name-giving processes of our villagers exist among American savage groups which reckon descent in the male line, and that they also existed among the savage groups which reckoned descent in the female line is, surely, a not unreasonable surmise. I add a list in parallel columns.

English Village Names	Siouan Group Names
Rogues	Bad Sorts
Stags	Elk[234]
Bull Dogs	Common Dogs
Horse Heads	Warts on Horses' Legs
Bone Pickers	Hide Scrapers
Pretty Girls	Girl Folk
Eaters of	Eaters of
Whitpot	Dried Venison
Cheese	Fish
Barley Bread	Dung
Dog.	

THEORY THAT SIOUAN GENTES NAMES ARE OF EUROPEAN ORIGIN

To produce, from North America, examples of group names conferred from without, as in the instances of our English villages, may, to some students, seem inadequate evidence. For example an unconvinced critic may say that the nicknames of Mr. Dorsey's 'Siouan gentes' were originally given by white men; the Sioux, Dacota, Asineboin, and other tribes having been long in contact with Europeans. Now it is quite possible that some of the names had this origin, as Mr. Dorsey himself observed. But no critic will go on to urge that the common totemic names which still designate many gentes were imposed by Europeans who came from English villages of 'Mice,' 'Cuckoos,' 'Tater Grubs,' 'Dogs,' and so forth. We might as

---

[234] Many other animal and vegetable names—totem names in America, village names in England—have already been cited. See p. 170.

wisely say that our peasantry borrowed these village names from what they had read about totem names in Cooper's novels. To name individuals, or groups, after animals, is certainly a natural tendency of the mind, whether in savage or civilised society.

If we take the famous Mandan tribe, now reckoning descent in the male line, but with undeniable survivals of descent in the female line, we find that the gentes are:

Wolf	Bear	Prairie Chicken	Good Knife
Eagle		Flat Head	High Village

Here, out of seven gentes, four names are totemic; one is a name of locality, 'High Village,' not a possible name in pristine nomadic society. While there are hundreds of such cases, we cannot reasonably regard the American group nicknames as generally of European origin. Still more does this theory fail us in the case of Melanesia, where contact with Europeans is recent and relatively slight. Among such tribes as the Mandans, and other Siouan peoples, we see Totemism with exogamy and female kinship waning, while kinship, recognised by male descent, plus settled conditions, brings in local names for gentes, and tends to cause the substitution of local names and nicknames for the totem group name. Precisely the same phenomena meet us, as we are to see, in Melanesia.

CHAPTER IX

THE MELANESIAN SYSTEMS

We have, fortunately, an opportunity in Melanesia of studying, as it seems, the Australian marriage system in a state of decay.[235] The institutions of Melanesia bear every note of being Australian institutions, decadent, dislocated, contaminated and partially obliterated. Starting from New Guinea, we find a long archipelago sloping down, away from the east side of Australia,

---

[235] Mr. Haddon agrees on this point.

towards the Fiji Islands. The archipelago consists mainly, in the order given, of New Ireland, New Britain, the Solomon Group, Banks Island, the New Hebrides, Loyalty Island, and New Caledonia. The inhabitants are a fusion of many oceanic elements, and are much more advanced in culture than the natives of Australia: they have chiefs, whose office tends to be hereditary (and in one place, Saa, is hereditary), in the male line, the father handing on to the son his magical acquirements and properties, and leaving to him his wealth, as far as he may. This is not very far, as, curious to say, descent in the female line is generally prevalent. Wealth is both real and personal: landed property consisting (1) of Town Lots, (2) of Gardens (ἕρκος), (3) of the Waste ('the Bush'). The 'town lots' and gardens pass by inheritance; the possessor being only 'possessor,' not proprietor, and real property passing in the female line, where that line still prevails. The reclaiming of land from the Waste tends, however, to direct property into the male line, which, except in certain districts, is not dominant. Money is divided, on a death, among brothers, nephews—and sons, 'if they can get it'—the money being the native shell currency. The tendency towards the substitution, as heirs, of a man's sons for his sister's sons, is powerful.[236]

This is a curious and anomalous condition of the family. As regards material advantages (χορηγία, Greek [choreigia]) Melanesian society is greatly in advance of Australian. It is in possession of houses, fruit trees, agricultural allotments, domesticated animals, and a native currency. Thus there is much property to be inherited, and where that is the case, and where the family has a house of its own, the desire of men to leave their goods and dwellings to their sons usually results in the reckoning of descent on the sword side. Yet, in this respect, the Melanesians of many regions are behind the naked, houseless Arunta, and other Australian tribes with male descent.. What influences caused these tribes to depart from the reckoning in the female line, still used among their equally destitute neighbours, the Urabunna, is a most difficult question; indeed the number of distinct grades, in relation to family laws among the Australians, is an enigma. Among the Melanesians, at all events, material advance and accumulation of property have often failed to bring inheritance out of the female into the male line.

Insular conditions are apt to develop divergences from any given type—local varieties—while the mixture of races, and the introduction into one island, or part of it, of the customs of settlers

---

[236] Codrington, The Melanesians, chaps, iii. iv.

from other islands, produces peculiarities and anomalies in Melanesia. We expect, therefore, to find Melanesian marriage rules rather dislocated and contaminated, and to see that the archaic type is half obliterated. In fact, this is the case, and Totemism, if it exists, survives in fragments and vestiges.

'Where are the totems?' Dr. Codrington asks, and we can only reply that they seem to be half obliterated. 'Nothing is more fundamental than the division of the people into two or more classes, which are exogamous, and in which descent is counted through the women.'[237] This answers to the Australian 'primary divisions,' or 'phratries.' But, in Australia, as we showed, these divisions appear to be of totemic origin. If this was so, in Melanesia, the evidence for the fact is much less distinct. In a large region of the Solomon Islands 'there is no division of the people into kindreds, as elsewhere, and descent follows the father.... The particular or local causes which have brought this exceptional state of things are unknown.'[238]

Speaking generally, however, the two primary exogamous classes exist, and to a Melanesian man, all women of his own generation count either as 'sisters' (barred) or as (potential) 'wives.' The appropriation of actual wives to their actual husbands 'has by no means so strong a hold on native society,' as the exogamous class divisions. By many students this license will be considered a survival of 'group marriage.' Prenuptial unchastity is wrong, but a breach of the exogamous rule used to be punished by death. Wife-lending used to be common, as in Central Australia, if the wife and guest were of opposite 'divisions.' Whether the license of certain feasts (as among Australians and Fijians) smiles on breaches of the exogamous law, does not seem quite certain.[239]

In Banks Island and the North New Hebrides, there are but the two 'primary class divisions.' These have not names as in Australia—if once they had names, the names are lost. We find merely 'divisions' (veve), two 'sides of the house.' Every man knows his own division; all the women in it are tabu to him; all the women of the other division, in the same generation, are potential wives (with certain restrictions in practice).

In Merlav, one of the Banks Islands, there are 'families within the kin' (answering to gentes—totem kins—in Australia). These families have local names, as a rule: one has its name from the Octopus, but eats it freely.

---

[237] Op. cit. p. 21.
[238] Op. cit. p. 22.
[239] Ibid. p. 26.

It is not inconceivable that here we have broken down and obliterated Totemism, among a settled agricultural people, probably dwelling, as a rule, in close contiguity.

In Florida, and adjacent parts of the Solomon Islands, not merely two, but six 'kema' or exogamous divisions ('phratries?') exist. Two of the six have names derived from localities, two have animal names, Eagle and Crab: two kema came in from abroad. All this points to contamination, and rearrangement, under new circumstances. Each kema in Florida has one or more buto, the clam, pig, pigeon, and so on, not to be eaten by members of the kema. This looks like the 'totemic subdivisions' (that is, the totem groups within the 'phratries') of the Australians. Again, these butos within each kema, animals and plants not to be eaten, are exactly like the survivals of Totemism in the names of the Siouan totem kins with male descent, 'Do not eat small Birds,' 'Do not eat Dogs,' 'Do not eat Buffalo,' and so forth. The buto of each kin within the Melanesian exogamous kemas, then, seems to me to be the old totem of the kin, now relegated to a position more obscure, in the changes of society, and, with one exception, not giving its name and tabu to the kema. Only in one case is the animal which is the buto, also the animal which gives its name to the kema. The Kakau kema may not eat Kakau—the crab. The Manukuma (eagle) kema may eat the eagle: one fancies that they find it tough. In the same way the Narrinyeri and other tribes in Australia permit their totem kins to eat their totems. Members of each kema are apt to speak of their butos (which they may not eat) as their ancestors, as in Totemism, but this is a mere mythical explanation of why they may not eat the buto. With half a dozen other myths, it is used by totemists to explain why they may not eat their totems.

Dr. Codrington, on the other hand, writes, 'the buto of each kema is probably comparatively recent in Florida, it has been introduced at Bugotu within the memory of living men.'[240] Dr. Codrington, as we have already seen, inclines to the theory which derives totems originally from individuals. He cites Mr. Sleigh, of Lifu (mentioned by us before), who writes, 'When a father was about to die, surrounded by members of his family, he might say what animal he will be, say a butterfly or some kind of bird. That creature would be sacred to his family, who would not injure or kill it; on seeing or falling in with such a creature the person would say, "That is kaka" (papa), and would, if possible, offer him a young cocoa-nut. But they did not thus adopt the name of a tribe.'[241]

---

[240] Op. cit. p. 32.
[241] Tylor, J. A. I., August, November. 1898, p. 147.

We need not repeat the objections to all such theories of the derivation of pristine totem group names from individuals, The butos, ancestors, not to be eaten, have all the air of archaic totems, now reduced to a lower plane, and, save in one case out of six, not giving the name to a kema, in Florida. Thus the butos of each kema would be, originally, totemic, but immigrations, settled conditions, the tendency to male descent, and the introduction of local or place names for some groups, of nicknames for others, broke down the old totemic nomenclature, leaving only the Kakau, or crabs, true to their colours and to their totem and totem name, while the other kemas got local names or nicknames—the Hongokikki being named from the pastime of Cat's Cradle—clearly a nickname. Apparently the pigeon is their buto. How did these conditions arise?

Say that there were once four exogamous totem groups in Ettrickdale—Grouse, Deer, Hares, Partridges. Say that there came in two alien groups, Trout and Plover. Of these two, one might come to be called Quoits, from their skill in that game.—Two of the original four might get local names, from their places of residence, say Singlee and Tushielaw. One might keep its old totem, Grouse, and its old totem name, abstaining from grouse. One might get a new name, Roe Deer, but all, under the names of Tushielaw, Singlee, Quoits, Roe Deer, and Grouse (with another not given), would retain their old totems as butos, ancestral in some way, and not to be eaten. But the new, not the totemic, names would now mark off the exogamous kemas. Something of this kind must have occurred in Florida, under new social conditions, and the stress of immigrants. But Dr. Codrington gives a case in which the banana was tabued, just before his death, by 'a man of much influence who said that he would be in the banana.'[242]

This origin of Totemism (namely, in animism, a man of influence tabuing, and bequeathing to his descendants for ever, the animal or plant that is to be his soul vehicle) is approved of, as the original cause of Totemism, by Dr. Wilken. But could it arise in a much lower state of society, wherein 'men of much influence' are rare, and are readily forgotten? Now in Melanesia, generally, a man's fame, however great, perishes with those who remember him in his life.[243] Again, this sort of tabuing the banana affected 'all the people' of the isle Ulawa, and so could not be the base of an exogamous prohibition, unless all brides were to be brought in from foreign islands. If the prohibition was confined to known

[242] Op. cit. p. 33.
[243] Ibid. p. 40.

143

descendants of the banana man, then we have the patriarchal family, founded by a known ancestor, and exogamous. Now, in Ulawa, descent is reckoned in the male line, and there are no exogamous divisions.[244] 'This is an exceptional state of things,' says Dr. Codrington (p. 22), yet he thinks it (p. 32) 'in all probability'—plus the tabuing of an object by a dying patriarch—the cause of the buto prohibition in the kemas of Florida. Thus a solitary case from an isle without exogamous divisions ('the only restriction on marriage is nearness in blood'), and with male descent, is supposed by Dr. Codrington to cause the buto prohibition in an island with exogamous divisions, and with female descent.[245]

His theory is manifestly inconsistent with his facts—moreover, it involves the existence of the patriarchal system at the time when totems first arose.

On the whole, this reasoning does not convince, but, if Dr. Codrington is right, Melanesian institutions are shattered, dislocated, contaminated, and worn down to a remarkable degree. Yet, behind them, where the two, or the six exogamous divisions prevail, with descent counted in the female line, we can scarcely help recognising a basis of Australian customary law, with obsolescence of the totem, slowly tending towards inheritance through the father. 'A chief's sons are none of them of his own kin; and, as will be shown, he passes on what he can of his property and authority to them.'[246] In spite of the 'generation names,' 'father,' 'brothers and sisters,' 'children,' the real distinctions of own father, cousin, and so forth, are understood, and expressed, as they usually are, everywhere.[247]

Thus Melanesia shows us some of the ways out from Totemism, exogamy, and descent in the female line. It also shows us, what Australia does not, ghost worship: most prominently in Saa, where, with descent in the male line, and hereditary chiefship, eleven generations of ancestors are remembered, 'by the invocation of their successive names in sacrifices.'[248] This is a solitary case of such genealogical knowledge among Melanesians, as distinct from Polynesians. It is made possible by the sacrifices to the ancestors. Now, in Australia, there are no such sacrifices. Without them ancestors among low savages cannot be remembered, and could not

---

[244] Ibid. p. 22.
[245] Dr. Codrington's exact words are 'The buto is in all probability a form of the custom which prevails in Ulawa,' and the banana story follows.
[246] Op. cit. pp. 33, 59-68.
[247] Ibid. pp. 36-37.
[248] Ibid. p. 50.

hand down, as an hereditary totem, the animal or other object which is their 'soul-box,' or the vehicle of the ancestral soul after death. There appears to myself to exist, in Melanesia, a notable tendency to adore, nay, almost to deify, a dead man, as a tindalo. Dr. Codrington cites, from Bishop Selwyn, a case in which a renowned brave man was slain in action. A house, or shrine, was built over his head, and he was canonised, or made a tindalo.

His claims to sanctity were automatically certified by canoe tilting, in principle like our table tilting. The men in the canoe cease paddling, 'in a quiet place,' and, when the canoe begins to tilt, they call over a roll of names of tindalos (human ghosts). At the name of the dead warrior,'the canoe shook again.' A successful raid followed, a new shrine was built for the warrior, and fish and food were sacrificed to him. By this means a great man's memory is, now and then, contrary to general custom, kept green in this region of Melanesia. Occasionally he seems to be on the way towards godship, as a departmental deity, perhaps as god of war.[249] Pigs are common victims, now, in sacrifice. We do not hear of any 'totem sacrifice,' if ever such a thing anywhere existed. In the case of a tindalo called Manoga, deification seems close at hand. His 'dwelling is the light of setting suns,' or of the dawn: or in high heaven, or in the Pleiades, or Orion's belt. It is a remarkable circumstance that this discarnate spirit is the tindalo or saint of a kema, or exogamous division, one of the six of Florida, and all of the six possess their tindalo, a ghost patron in receipt of sacrifice, as well as their buto, or animal not to be eaten.[250]

Still more remarkable it is that, in certain Melanesian isles of the New Britain group, the two exogamous divisions are neither anonymous, nor totemic, nor of local names, nor bear nicknames, but are named after the two opposing powers of Dualism, the God and Devil of savage theology. Of these Te Kabinana is 'the founder, creator, or inventor of all good and useful things, usages, and institutions.' On the other hand To Kovuvura is the Epimetheus of this savage Prometheus: Te Kabinana created good land: To Kovuvura created bad land, mountains and everything clumsy and ill formed. These powers captain the two exogamous divisions, an office assumed by two totems in the neighbouring Duke of York group.[251] Nothing can prove more clearly the blending of different stages of thought in Melanesia.

On the whole, Totemism is breaking down, and something

---

[249] Op. cit. pp. 124-130.
[250] Danks, J. A. I. xviii. 3, 281-282.
[251] Ibid. pp. 131-132.

very like Polytheism, of an animistic type, is beginning to emerge, in Melanesia. There is a tindalo of the sea, of war, and of gardens,— Poseidon, Ares, and Priapus in the making. Sacrifice and prayer exist, neither is found (perhaps with an exception as regards prayers for the souls of the dead) in Australia. On the other hand, only the smallest of small change for the Australian conception of such makers and judges as Baiame is noted in Melanesia, mainly in the myths of and prayers to Qat, and myths of a creative unworshipped female being. These are Vuis, not ghosts; they are spirits never incarnate, unlike the tindalos.[252] Qat appears to hover between the estate of a lowly creative being, born of a rock, and that of a culture hero, and rather resembles the Zulu Unkulunkulu. Thus Melanesia seems, in society and beliefs, to show an advance from Totemism, nomadic life, and from an unworshipped female creative being, towards Animism and Polytheism, and descent reckoned in the male line: agricultural and settled existence, with mixture of race, and foreign contamination of custom, being marked agents in the developement.

As tindalos (human ghosts, in one case the patron of a kema) thrive to Gods' estate, while butos remain ancestral plants or animals, not to be eaten, it would be a natural step to imagine later that the family God (tindalo) of ghost origin, incarnates himself in the buto, the sacred animal of the kin. That would be an explanatory myth. If accepted, it would produce the Samoan and Fijian belief, that the animals and plants not to be eaten by the kindreds (old totems) are incarnations of gods. Thus the Florida beliefs and customs are a stage between those of Australia and those of Samoa and Fiji.

HOW THE ORIGIN OF TOTEM NAMES WAS FORGOTTEN

It appears, at least to the mind of the maker of an hypothesis, that the names of Melanesian kemas, as well as the new names of American 'gentes' (totem kins with male descent), indicate the probability that, from the first—as among our villagers—group names were given (in the majority of cases) from without, as in many American and some Melanesian cases they certainly are. We see that it is so: no group would call itself 'Cat's Cradle Players,' or 'Eaters of Hide-scrapings,' or 'Bone Pickers,' as in Florida; among the Sioux; and in Western England. We cannot possibly expect to

---

[252] Codrington, op. cit. pp. 154-156.

find any groups in the process of becoming totemic and of having plant and animal names given to them from without. But we certainly do observe that names, or nicknames, relatively recent, are given to savage groups, on their way out of Totemism—the totem name often still lingering on in America, like the butos in Melanesia—and that these names, or nicknames, are given from without. Nearer to demonstration that the totem names were given in the same way (as 'Whig' and 'Tory' were given), we cannot expect to come.

It may be said that my conjecture is only a form of that suggested (if I understand him) by Mr. Herbert Spencer. An individual had an animal name or nickname. He died: his ghost was revered by his old name, say Bear. He was forgotten, and his descendants, who kept up his worship, came to think that they were descended from a real bear, and were akin to bears. I need not once more reiterate the objections to this theory, but, like my own suggestion, it involves forgetfulness of a fact,—here the fact that 'Bear' was a human ancestor. Against the chances of this forgetfulness was the circumstance that individuals were constantly being named Bear, Wolf, Eagle, and so on, in daily experience, usually with a qualifying epithet, 'Sitting Bull,' 'Howling Wolf,' and so forth. These facts might have prevented Mr. Spencer's savages from forgetting that the ancestral Bear was a Bear of human kind, like themselves and their contemporaries.

In my hypothesis, forgetfulness, on the other hand, might readily occur. When all the group names in each area had become organised and stereotyped, there would necessarily be no new giving of group names to remind the savages how these titles came into existence. On the other hand the myth-making stage, as to kinship with the name-giving plants and animals, would set in, and then would come reverent behaviour towards these creatures, as if they were kinsmen and friends. Respect for the totem, in each case, will clinch the tendency to group exogamy. I have supposed, for the reasons given, that there was already a tendency against marriage within the group. That tendency must have been confirmed by the totem tabu against making any use of any member of the totem kin, and a woman of the totem would be exempted from marital use by her male fellow-totemists. The totem belief would add a supernormal sanction to the exogamous tendency.

OTHER SOURCES OF SACREDNESS IN PLANTS AND ANIMALS

Now any such superstitious respect for an animal, whatever its origin, will take the same inevitable forms; and thus, if individuals select nyarongs, naguals, and so on, they must necessarily behave to these things as they do to their hereditary totems. There is no other way in which they can behave, if they regard the animals as mysteriously friendly and protective, though the idea that they are friendly and protective has different origins, in either case.

Thus the exigencies of my guess as to the origin of Totemism, compel me to disagree with a dictum of Mr. Frazer, 'if the relations are similar, the explanation which holds good of the one ought equally to hold good of the other.'[253] The conclusion is not necessary. You may revere a rat (your totem), and a cat (your nagual) for quite different reasons, and in quite different capacities, you being the kinsman of your totem, the protégé of your nagual; but, if you revere them, your reverence can only show itself in the same ways. There are no other ways.[254]

RECAPITULATION

Does my guess at the origin of totems seem out of harmony with human nature? You, belonging to a local group, must call other groups by one name or another. Plant and animal names come very handy. The names fluctuate at first, but are at last accepted by the groups to which they are applied. The origin of the names being

---

[253] Golden Bough, iii. 416-417.

[254] Mrs. Langloh Parker writes, concerning the Euahlayi Baiame-worshipping tribe of New South Wales: 'A person has often a second or individual totem of his name, not hereditary, and given him by the wirreenuns' (medicine men), 'called his yunbeai, any hurt to which injures him, and which he may never eat—his hereditary totem he may. He is supposed to be able, if he be a great wirreenun, to take the form of his yunbeai, which will also give him assistance in time of trouble or danger, is a sort of alter ego, as it were.' In this tribe the yunbeai (nyarong, nagual, manitu) is of more importance to the individual than his hereditary totem, which, however, by Baiame's law, regulates marriage, as elsewhere (Folk-Lore, x. 491, 492). The tribe studied by Mrs. Langloh Parker speaks a dialect (Euahlayi) akin to the Kamilaroi, but the Kamilaroi of Mr. Ridley are seated three or four hundred miles away.

forgotten, an explanation of them is needed, and, as in every case where it is needed, it is provided in myths. The myths, once believed in, are acted upon; they become the parents of tabus, magic, rites of various kinds. Social rules must be developed, some already exist; and each group called by an animal, plant, or other such name, becomes, under that name, a social unit, and accepts, as such, the customary legislation, just as a parish does. You must not marry within the totem name: either because of the totem tabu in general, or because the totem comes to be conceived of as denoting kinship, and (for one reason or another) you had already a tendency not to marry within the limit of the group. The usual totem rules may be thwarted by other rules derived from a peculiar system of animism, very philosophically elaborated, as among the Arunta of Central Australia. The institution, in short, may develop or may dwindle, may persist in practice, or fade into faint survival, or blend with analogous superstitions, or wholly vanish, in varying conditions. Totemism affects art; to some extent it may have affected religious evolution. It is certainly a source of innumerable myths.

But, if my guess holds water, Totemism arose out of names given from without, these names being of a serviceable sort, as they could be, and are, not only readily expressed in words, but readily conveyed in gesture language from a considerable distance. The names could be 'signalled.'[255] 'There is an Emu man: look out!' This could easily and silently be expressed in gesture language. Place-names, and many nicknames, could not so be signalled.

This theory, of course, is not in accordance with any savage explanations of the origin of their totem. It could not be! Their explanations are such fables as only men in their intellectual condition could invent: they are myths, they involve impossibilities. My hypothesis (or myth) does not, I think, involve anything impossible or far-fetched, or incapable of proof in a general way. It is human, it is inevitable, that plant and animal names should be given, especially among groups more or less hostile. We call the French 'frogs.' It is also a fact that names given from without come to be accepted. It is a fact that names, once accepted, are explained by myths; it is a fact that myths come to be believed, and that belief influences behaviour.

---

[255] Roth, Ethnological Studies, 71-90. Dr. Roth gives the signs for the animals, but does not say that they are used for signalling totem names; indeed, he says nothing about totems.

AN OBJECTION ANSWERED

Here I foresee an objection; it will be said that, on the other hand, behaviour produces myths. Men find themselves performing some apparently idiotic rite: they ask themselves, 'Why do we do this thing?' and they invent a myth as an answer. Certainly they do, but you believe in a God, or in Saints, and act (or you ought to act) in a manner pleasing to these guardians of conduct. You don't believe in a God, because you behave well, and it is not because you behave well to a totem that you believe in a totem. You treat him as game, not as vermin, because you believe in him, and your belief is based on the myth which your ancestors invented to account for their having a totem.

My guess has the advantage of going behind the age of settled dwellings, agriculture, kinship through males, and the causal action of individuals. It reverts to the group stage of human life. Groups give and accept the names; invent the myths, act on their belief in the myths, and so introduce the sanction of what had perhaps been a mere tendency towards exogamy. On the other hand, my guess has the disadvantage of dealing with a hypothetical stage of society, behind experience. But this cannot be avoided, for if we base our hypotheses of the origin of Totemism on our experience of the ways of societies which have passed, or are passing, out of Totemism, our theories must necessarily be invalidated. It may be replied that I have myself given illustrations of my theory from the folk-lore of civilised society. But the only begetters of these illustrative cases are boys—and boys are in the savage stage, 'at least as far as they are able.'

In a tone more serious, it may be reiterated that no theory of the origin of Totemism is likely to be correct which derives the totem, in the first instance, in any way, from the individual, the private man. Long ago, Mr. Fison wrote, 'Sir John Lubbock considers that the "worship of plants and animals is susceptible of a very simple explanation, and has really originated from the practice of naming, first individuals, and then their families, after particular animals."'[256] Mr. Fison replied, 'This is surely a reversal of the true order. The Australian divisions show that the totem is, in the first place, the badge of a group, not of an individual. The individual takes it, in common with his fellows, only because he is a member of the group. And, even if it were first given to an individual, his

---

[256] Origin of Civilisation, p. 183.

family, i.e. his children, could not inherit it from him,' when descent is reckoned on the female side.[257]

It is a commonplace, perhaps an overworked commonplace, that the group, not the individual, is the earlier social unit. Yet the hypotheses of Lord Avebury, Mr. Herbert Spencer, Dr. Wilken, Mr. Boas, Miss Alice Fletcher, and Messrs. Hose and McDougall, all derive from individuals, in one way or other, the most archaic names of human groups. The hypothesis of Mr. Max Müller leaves the origin of the group name unexplained. The later hypothesis (especially provisional), of Mr. Frazer, does start from the group name, but I am not certain whether we are to understand that each group name is derived from the plant Or animal or selected by the group as the object of its magical rites, or whether, for some unknown reason, each group already bore the name of the animal, or plant, or element, before entering on the great co-operative industrial system. Now it seems to me certain that the names, in each case, were originally not names of individuals, or in any way derived from individuals, but were names of groups. As to how pristine groups might obtain such names I have offered what, in the nature of the case, has to be only a conjecture. But named, as soon as men had intelligence and speech, the groups, as groups, had to be, and the actual names are such as, whether in savagery or in full civilisation, are given to individuals, and are also, in civilised rural society, given to local groups, to members of parishes and villages. So far, the cause which I suggest is a vera causa of collective group names.

OTHER OBJECTIONS ANSWERED

A well-known Folk-lorist to whom I submitted my theory, rather 'in the rough,' replied to me thus: 'I have thought of Totemism as meaning a social system, that is, as including belief, worship, kinship, society. And therefore, the animal or plant names are an essential part of the system. You, as I understand it, come along and say the name is the result of one of the trifles of the human mind, therefore did not enter into the totem system very deeply, and certainly did not belong to the beliefs and the worship, except as the result of a later myth-making age. Of course your book may explain all, and I shall look forward to studying it, as I have always enjoyed your studies.

---

[257] Kamilaroi and Kurnai, p. 165. In his edition of 1902, Lord Avebury does not reply to these arguments.

'But I confess I don't much believe in these accidents causing or rather entering into so widely spread a system as Totemism. Cut away the name and nothing is left to Totemism except myth, survivals, and a social grouping without any apparent cement. Blood kinship as a basis of society surely arose much later, unless Dr. Reevers's remarkable evidence from the Haddon expedition to New Guinea helps the matter. He found, you remember, blood kinship traceable by definite genealogies beneath, so to speak, a system of Totemism, and but for the most minute examination blood kinship would have escaped observation once more and Totemism only would have been reported. Is this blood kinship the true social basis and Totemism only a veneer?

'I have goodly notes on Totemism and non-Totemism, and I confess it difficult to eliminate the name as an important part of the system. It covers every part—is the shell into which all the rest fits. Now I have too much respect for our savage friends to think they used myth any further than we do. We go every Sunday saying "I believe," but we don't build up much upon this. Our social fabric, nay our religion, is not of this. And so of the savage. If I grant you the myth of descent from an animal to have arisen out of a pre-existing name system, I am no nearer the understanding of totem-kinship as the basis of a social group.'

These are natural objections, on a first view of my suggestions. Totemism is a social system, but there was an age before totemism, an age of undeveloped totemism; into these we try to peer. But the method of name-giving which I postulate is hardly 'a trifle of the human mind.' It is, as I have proved, a widely diffused, probably an universal tendency of the human mind. Not less universal, in the savage intellectual condition, is the belief in the personality and human characteristics of all things whatsoever; man is only one tribe in the cosmic kinship, and is capable of specially close kinship with animals. Nobody denies this, and the resulting myths to explain the connection of the groups and their totems are not only natural, but inevitable—the real origin of the connection, 'in the dark backward and abysm of time,' being forgotten. We may go to church, and say 'I believe,' and we may not act up to our creeds. 'And so of the savage.' But it is not 'so of the savage.' His belief in a myth of kinship with an Emu is carried into practice, and regulates his conduct, magical and social. This is not contestable. In the same way a Christian who believes in the efficacy of masses for the welfare of his dead friends, pays for masses. At the lowest, he 'thinks the experiment well worth trying.' To other myths, say as to the origin of the spots on a beast, a savage may 'give but a doubtsome credit.' They are not of a nature to affect his conduct in

152

any way. But the totem myths do affect his conduct, quite undeniably, and, even if there are sceptics, public opinion and customary law compel them to regulate their behaviour on the lines of the general belief. We are not to be told that nobody believes in anything! The 'social grouping' consequent on the beliefs is not 'without any apparent cement.' The cement is the belief in the actual kinship of all persons having the same totem name, and sacred totem blood, even if they belong to remote and hostile tribes. All wolves are brethren in the wolf; all bears are brethren in the bear; and so men-bears are sisters to women-bears, and brothers may not marry sisters. Here is 'apparent cement' of the very best quality, and in abundance, given the acknowledged condition of the savage intellect.

Manifestly these ideas belong, as a whole, to 'a later myth-making age '—that is to an age later than the dateless period of the hypothetical anonymous groups. But, between that hypothetical period and the evolution of the idea of group kinship with animals and plants, and with all men of the same animal and plant stock names, there is time enough and to spare for the full evolution of Totemism.

Again, to a Darwinian, the enormous influence of 'accidents' in evolution ought not to be a matter hard of belief; without 'accidents' (in the Darwinian sense of the word), there would be no differentiation at all, and no evolution. The Darwinian 'accident' seems to mean a variation of unknown cause. But the giving of plant and animal group names is hardly an 'accident' of this kind. 'What else are you to call it?' the player asked, when questioned as to the origin of the words 'a yorker.' And by what names so handy and serviceable as plant and animal names were pristine men to call the neighbouring groups?

I have shown why place names were less handy, and how, in nomadic life, they were scarcely possible. Local names come in as Totemism goes out. Long nicknames, 'Boil-food-with-the-paunch-skin,' 'Take down their leggings,' 'Travel-with-very-light-baggage,' 'Shot-at-some-white-object' (Siouan nicknames of gentes), are much less handy, much less easy to be signalled by gesture language, and are certainly much later than 'Emu,' 'Wolf,' 'Kangaroo,' 'Eagle,' 'Skunk,' and other totem names. If such totem names were, originally, the favourite form of nomenclature for hostile groups (like our 'Sick Vulture' for a famous scholar, or 'Talking Potato,' for Mr. J. W. Croker), I see not much of an 'accident' in the circumstance.

The totem names, then, came in upon a very early society: and

153

myth, belief, custom, and rite, crystallised round them, and round the idea of blood kindred, which must be very early indeed.

My critic asks, 'Is blood kinship the true social basis, and Totemism only a veneer?' That question I have already answered. In my opinion mankind, in evolving prohibitions of marriage, first had their eyes on contiguity, that of 'hearth-mates.' Groups of hearth-mates were next distinguished by totem names. But these names could give no superstitious sanction to customary laws, till the idea of 'blood kinship' with, or descent from, or evolution out of, or other form of kinship with the totem was developed. At this period, the totem name roughly indicated ties of blood kinship. But the Australians, as we saw, have now reached a clearer idea of what blood kinship is, and, by a bye-law, prohibit marriages of 'too near flesh,' in cases where, though the persons are akin by blood, totem law does not interfere. Totem law has had an educating influence in developing the objection to marriages between people contiguous as hearth-mates, into the objection to marriages between persons too near in blood kinship. Thus Totemism is not 'only a veneer.'

On the foundation of all these blended ideas, Totemism arose, a stately but fantastic structure, varying in shape under changing conditions, like an iceberg in summer seas. It is, indeed, 'a far cry' from anonymous human groups, and groups of plant or animal names, to Helen, the daughter of the swan, that was Zeus! But the pedigree is hardly disputable.

On the other hand, suppress the totem names, give the original groups such titles as the Sioux 'Take-down-their-leggings,' or 'Boil-meat-in-the-paunch-skin' (some names you must give them), and what is left? Suppose such names to have been those of pristine groups, and suppose them to be tending to exogamy. A 'Boil-meat-in-the-paunch-skin' man may not marry a 'Boil-meat-in-the-paunch-skin' girl; but must marry a 'Take-down-their-leggings' girl, or a 'shoot-in-the-woods' girl, or a 'Do-not-split-the-body-of-a-buffalo-with-a-knife-but-cut -it-up-as-they-please' girl! That is rather cumbrous: marriage rules on that basis are not readily conceivable.

And where is here the tabu sanction? Brother Wolf or Brother Emu is a thinkable, powerful, sacred kinsman, who will not have his tabu tampered with. But there is no sanctity in Do-not-split-the-body-of-a-buffalo-with-a-knife-but -cut-it-up-as-they-please!

Luckily we have here a case in point. My theory is that animal names being once given to the groups, the animal, in accordance with savage ideas, became a kinsman and protector. The animal or vegetable or other type, in each case, sanctioned various tabus, including exogamy. Had the name been another kind of nickname,

154

as 'Boil-meat-in-the paunch-skin,' what was there to sanction the tabu? Or, if the group name was a local name, where was the sanction? Exogamy does persist where totem groups have become local, and are now known by the names of their places of settlement. But not always. Of an Australian tribe, the Gournditch Mara, we read that it consisted of four local divisions, water (mere?), swamp, mountain, and river. But there was no exogamous rule affecting marriage. A man of the group dwelling in the swamp might marry a woman of the same group. There was descent in the male line; wife-lending was highly condemned. The office of headman was hereditary in the male line, 'before any whites came into the country.' The benighted tribe was not devoid of superstition.

'They believed that there was a future good and bright place, to which those who were good went after death, and that there was a Man at that place who took care of the world and of all the people.' The place was called Mūmble-Mirring. The dark, bad place was Burreet Barrat. 'This belief they had before there was any white person in the country.'

As these statements are odious to most anthropologists, they cannot be true, and thus a slur is cast on all that we learn about the Gournditch Mara. But though a missionary (the Rev. Mr. Stähle) cannot, of course, be trusted here, he had no professional motive for fictions about the marriage laws of the tribe. They had no ceremonies of initiation, no seasons of license, apparently no totems, and the merely local names of groups naturally carried no exogamous prohibition: conveyed no tabu sanction.[258] Had there never been any totem names, exogamy might never have arisen.

How my friendly critic is 'no nearer to the understanding of totem kinship as the basis of a social group,' if, for the sake of the argument, he grants 'the myth of descent from an animal to have arisen from a pre-existing name system,' I am at a loss to comprehend. Here are groups, Bear, Wolf, Trout, Racoon, firmly, though erroneously, believing that they are akin to these animals. Naturally they 'behave as such.' Each racoon has duties to other racoons, and to the actual racoons. He does not shoot a racoon if he can get anything else; he does not shoot a racoon sitting. He is brother to racoons of his own sex, and to sisters in the racoon of the other sex. He does not marry them. The belief in the racoon kinship is the basis of that social group—the man has other social groups of other kinds. Savages believe in their beliefs, to the extent of dying from fear after infringing a tabu in which they believe. Thus I would reply to the objections offered after a first glance at my conjecture.

---

[258] Kamilaroi and Kurnai, pp. 274-278.

TOTEMS AND MAGICAL SOCIETIES

A man has other social groups than his own totem group in certain regions. Totem groups among the Arunta, we have seen, work magic 'to secure the increase of the plant or animal which gives its name to the totem.' The Arunta have no myth as to the origin of these performances, styled Intichiuma.[259] This, as far as Australia is concerned, seems to be a peculiarity of the Arunta system alone, or all but alone, and, as we saw, it has even been suggested that these rites are the origin of Totemism. But such rites appear to be most firmly established and organised among societies which are passing out of Totemism. Such a society is that of the Omaha tribe of North America, where descent is reckoned in the male line.[260] Among the Omahas we find the Elk totem group with male kinship; they may not touch a male Elk, or eat its flesh: if they do, as in New Caledonia, they break out into sores. This kindred, with the Bears, 'worship the thunder' in spring. Their special business and duty is 'to stop the rain.' But, if they are a Weather Society, in this respect, that fact does not appear in their totem names, Elk and Bear.

Other Omaha gentes, or 'subgentes,' are also totemic, and are named from that which they may not eat, as wild turkeys, wild geese, cranes, and blackbirds. The people of the black-bird totem actually do a little totem magic, against their totem; they chew and spit out corn, to prevent the blackbirds from feeding on the crops.[261] The reptile group does not touch or eat reptiles, but, if worms injure the corn, they pound a few worms up into flour, make a soup thereof, and eat it (is this 'totem sacrifice'?), all for the good of the crops. The worm group does this magic (involving the eating of its totem) not for the benefit of worms (as among the Arunta) but to control the mischievous action of worms.

Now turning to Magical or Magico-Religious Societies among these Indians, we find a Wind Society, but it contains members of many totems, buffalo, eagle, hawk, and so on, plus 'The South wind people,' who, apparently, may be a totem group of that name, which, as among the Arunta, might work wind-magic.[262] But our authority, the late Mr. Dorsey, calls all the members of this Wind Magic Society 'Wind gentes,' and surely this breeds much confusion.

---

[259] Spencer and Gillen, ch. vi.
[260] Dorsey, 'Omaha Sociology,' Bureau of Ethnology, 1881-1882, p. 225.
[261] Dorsey, 'Omaha Sociology,' Bureau of Ethnology, 1881-1882, pp. 238-239.
[262] Dorsey, 'Siouan Cults,' Bureau of Ethnology, 1889-1890 (1894), p. 537.

By a gens he usually means a totem kin with male descent (by 'clan,' he means a totem kin with female descent). Thus all 'wind gentes' ought to be wind totem groups: only wind totem groups ought to be in the Wind Society, which is not the case: and all water gentes, or earth, or fire gentes ought to be of water, earth, or fire totems. But this, again, is not the case.

All sorts of totem kindreds enter into the earth, wind, fire, and water Magical Societies, or Magico-Religious Societies. They belong to them as members of any universities, or of certain selected universities, may belong to an University Club: or, again, may be Catholics, Anglicans, Brownists, or Presbyterians. These American Magical Societies, though composed of members of totem kindreds, are not, in themselves, totemic societies. Members of other totems serve in the societies which work magic for earth, wind, fire, and so on. Among the Arunta, on the other hand, the magic for each object is worked solely by the men who have that object for totem. To a certain extent, however, this rule is changing, and members of other totems may, at least, be present at each totem's Intichiuma, or magical rites.[263] 'In addition to the members of the totem' (water) 'other men are invited to come, though they will not be allowed to take any part in the actual Intichiuma ceremony.' From presence, by invitation, to participation in the rites (as in the American Shamanistic Societies), is a step which may come to be taken, and thus the Arunta totem groups would become mere 'Shamanistic Societies.'

A most curious and interesting account of the Omaha Magical Societies is given by Miss Alice Fletcher, in her essay, already cited, on 'The Import of the Totem.' To obtain the 'personal totem' (manitu) a youth must first listen to his elders. They tell him 'to go forth to cry to Wa-kon-da. You shall not ask for any particular thing, whatever is good, that may Wa-kon-da give.'

Fiat voluntas tua!

'Four days and nights upon the hills the youth shall pray, crying, and, when he stops, shall wipe his tears with the palms of his hands, lift his wet hands to heaven, then lay them on the earth.'

To the ordinary mind, this describes such prayers as are the petitions of the Saints. But, in accordance with the views of the official school of American anthropology, it is averred that nothing of the kind is intended by the Omaha. 'There is no evidence that they did regard the power represented by that word (Wa-kon-da) as a supreme being, nor is there any intimation that they had ever conceived of a single great ruling spirit,' says Miss Fletcher (1897).

---

[263] Spencer and Gillen, pp. 169, 191.

The prayer is evidence enough. Prayer is directed to a person, and whether he is envisaged as 'a spirit,' or not, is a mere detail of metaphysical terminology. If Miss Fletcher is right, Wa-kon-da is a pantheistic conception, but as He, (or It) also listens to prayer, He (or It) is personal. We see rather an anthropomorphic conception of deity, passing towards pantheism, or to divinity no longer anthropomorphic, than a notion of impersonal force immanent in the universe, passing towards anthropomorphism—as in Miss Fletcher's theory. The idea of such a force, or cosmic rapport (the Maori mana), is, indeed, familiar to us in the speculations of the lower barbaric races. It does credit to their metaphysics, but, prima facie, seems likely to be later in evolution than the idea of an anthropomorphic Maker, like the Australian Baiame.

At all events, the Omaha appears to live, in prayer, on a high religious level, and it is open to the friends of religious borrowing, to say that he took his creed from Europeans. I am not certain that Miss Fletcher is indisposed to agree with me on this point of Red Indian unborrowed theism. In her Indian Song and Story,[264] she gives Pawnee songs, 'hitherto sealed from the knowledge of the white race.' Here is one:

> Lift thine eyes! 'Tis the gods who come near,
> Bringing thee joy, release from all pain.
> Sending sorrow and sighing
> Far from the child, Ti-ra-wa makes fain.
>
> Ah, you look, you know who comes,
> Claiming you his, and bidding you rise,
> Blithely smiling and happy,
> Child of Ti-ra-wa, Lord of the Skies!

Ti-ra-wa is Hau-ars, 'a contraction of the word meaning father.' The song is used to still children who cry at a religious ceremony.

However it be, the Omaha prays to Wa-kon-da, not for 'any particular thing,' but for whatever, in the gift of Wa-kon-da, is good, and mainly for a manitu ('personal totem'). The Omaha also believe in telepathy. 'Thought and will can be projected to help a friend.' A magical society exists, to concentrate and direct this expenditure of energy, and the process is strengthened by such things as the neophyte beholds in vision, after prayer to Wa-kon-da. He sent an answer to prayer, a feather of a bird, a tuft of a beast's hair, a crystal,

---

[264] Nutt, 1900, pp. 108-112.

a black stone, representing the species, or element of nature, which was to be the neophyte's 'personal totem,' or manitu. If it were thunder, the man could control the elements; if it were an eagle, he had an eagle eye for the future; if it were a bear (or a badger), he was not so gifted.

Now, according to Miss Fletcher, the Bear Magical Society is composed of men, who, after prayer, have seen the bear in dream or vision; those who saw representatives 'of thunder or water beings' form the Society which deals with the weather. 'The membership came from every kinship group' (totem kin) 'in the tribe.' Thus the Magical Societies are composed of men of any totem, and, the less purely totemic the tribe, the stronger is the Magical Society.

The totem kins now, among the Omaha, have descent in the male line. All this is 'late,' and 'late' is the totem priesthood held by 'hereditary chiefs of the gens.' Miss Fletcher regards the totem of the 'gens,' with the beliefs crystallised around it, as an ingenious 'expedient,' with a social 'purpose!' the totem of each kindred having been inherited from the vision and manitu of some ancestral chief. We need not again point out that, even now, among the Omaha, advanced as they are, manitus are not hereditable, and that Miss Fletcher's system cannot account for Totemism in tribes which reckon descent on the spindle side. Miss Fletcher justly remarks that the real totem, 'the gentile totem,' 'gave no immediate hold upon the supernatural, as did the individual totem' (manitu) 'to its possessor. It served solely as a mark of kinship, and its connection with the supernatural was manifest only in its punishment of the violation of tabu.'

In brief, the real totem, and the individual manitu, with its magical societies, are two things totally apart, and apart we must keep them, in our studies of early society. Not to do so is to make the topic incomprehensible.

TOTEM SURVIVALS

In other books, especially in Myth, Ritual, and Religion, and Custom and Myth, I have examined apparent survivals of Totemism, in ancient Greece, ancient Egypt, and other civilised countries. Of these the most notable are the Greek myths of descent of families from animals, explained as the temporary vehicles of Zeus or Apollo: and the worship of special animals by each of the Nomes of Egypt. Other arguments I have offered, especially in the case of Apollo and the Shrew Mouse. I remain of the opinion that many of the Greek mythical and religious phenomena noted, are most probably to be explained as survivals of a totemic past. Of course Totemism is only one element in animal worship, and the Corn Spirit, disguised as almost any animal you please, may be one of the other elements. But, as far as I have studied the subject, I agree with Mr. Tylor in his 'protest against the manner in which totems have been placed almost at the foundation of religion. Totemism ... has been exaggerated out of proportion to its real theological magnitude.... The rise and growth of ideas of deity, a branch of knowledge requiring the largest range of information and the greatest care in inference, cannot, I hold, be judged on the basis of a section of theology of secondary importance—namely, animal worship—much less of a special section of that—namely, the association of a species of animals (and of a vast variety of other things) 'with a clan of men which results in Totemism. A theoretical structure has been raised quite too wide and high for such a foundation.'[265] The totem god himself I regard as only the hypothesis by which certain barbaric races account to themselves for the survivals of Totemism among them. The so-called 'totem sacrament' is not 'god-eating,' but a piece of magic, used in ceremonies designed to foster—or to vex and annoy—the totem. As Mr. Tylor writes, 'till the totem sacrament is vouched for by some more real proof, it had better fall out of speculative theology.'

DID THE ANCESTORS OF THE CIVILISED RACES PASS THROUGH THE AUSTRALIAN STAGE?

That the ancestors of the Aryan-speaking peoples passed through the 'stone age' of culture, few will deny. That they also passed through the totemic stage as regards marriage law is, however, a problem perhaps not to be solved. For reasons unknown, the 'white' races (not to speak of Egyptians, Babylonians, Chinese, and Japanese) have a peculiar aptitude for civilisation, are peculiarly accessible to ideas. It might therefore be argued that conceivably they were readily accessible to the idea of blood kinship. The maternal affection, in a race whose children (unlike the offspring of the lower animals) are so long in attaining maturity, cannot but suggest the idea of blood kinship. Among totemic peoples it seems that this idea was originally defined by the totem name, a definition at once too wide and too narrow. It is not physically unthinkable that our own ancestors may have been more acutely intelligent, and, if so, why should not they simply forbid unions between persons too near akin in blood? We have found no

---

[265] J. A. I., August, November, 1898, p. 144.

such moral or instinctive reason among totemic peoples who were, apparently, led to exogamy, first by non-moral causes, or causes in which the moral element was not explicit, and then, by aid of corollaries from rules thus based, came to forbid marriages of 'too near flesh.' Without the training of totemic institutions, it is hard to see how the Aryan-speaking peoples (however naturally gifted from the first) arrived at the same conception of incest. It seems absurd to suppose that black men and red men arrived at the idea of incest, and at the laws which prohibit it, by the devious and unpromising path of Totemism, while white men reached the same point in some other way. Yet if it has appeared difficult to find traces of Totemism among the Melanesians, much more difficult must it be to prove that races with so long a civilised history as, for example, the Greeks, were once under totemic institutions.

I have already indicated my inclination to believe that Totemism has left its traces, in Greece, in the myths of descent from bulls, bears, swans, dogs, ants, and so forth, and in certain peculiar aspects of animal worship. It is usual for scholars to explain these facts away, as things borrowed by early Greeks from some other race. But 'the receiver is as bad as the thief,' and if Greeks were capable of accepting totemic ideas, they were capable of evolving totemic institutions. We are not to invent an ideal 'Aryan,' and then to explain all his traces of savagery as borrowings by him from some unknown prior race. There is no reason at all for supposing that the peoples who speak languages called, for convenience, 'Aryan,' were better bred than any other peoples at the beginning.

It would greatly add to the force of the presumptions in favour of an 'Aryan' totemic past, if we could point to apparent survivals not only in myth and early art, but in actual institutions. Now there are Greek institutions, in Attica, the 'deme,' the genos, and the phratria, which may be interpreted, rightly or wrongly, as survivals of Totemism. We have seen that gens (equivalent to the Greek γένος [genos]) and that phratria (Φρατρία) are used, by certain students, to designate the totem kin, and the two 'primary exogamous divisions' (say Dilbi and Kupathin) of Australia and North America. To use gens thus is misleading, especially as 'totem kin' is adequate and unambiguous. But we have here employed 'phratria' to designate the 'primary exogamous division,' because no better word is handy, while we do not maintain that the Attic phratria is a survival of the institution usual in Australia.

Messrs. Fison and Howitt, in an instructive paper, have offered, as a provisional hypothesis, the theory that the Attic deme (a local association) may have arisen from the kind of local tribe (or horde) in Australia, while the Attic phratries and γένη (associations

161

depending on birth and kinship) were survivals of the 'primary exogamous divisions' and totem kins.[266] The present writer had made similar suggestions long ago.[267] Concerning the γένος and Φρατρία we know but little: inevitably, for we have seen that, even in Australia, still more in Melanesia, local names and local communities are beginning to encroach on and usurp the authority of the totem kin, and other associations based on common blood, real or reputed. Infinitely more must this have been the case in Greece. If savage phratries and totem kins once existed in Attica, they must have been nearly obliterated long before the historical period. At most they would only survive in connection with ritual and religion. Again, our definitions of γένος and Φρατρία are derived from late grammarians and lexicographers. Thus our means of knowledge are limited and darkling.

Messrs. Howitt and Fison start from the horde, or tribe, the horde meaning the largest local Australian community, composed of subtribes, if we are not merely to say 'tribe,' and leave 'horde' out of the question. The members of the horde or tribe are, as we know, of many various totems, but of only two 'primary exogamous divisions' or phratries. Into these the members are born, mostly taking the mother's phratry and totem. As a rule, both father and mother belong to the tribe, but if a woman does come in out of an alien tribe, her children, though deriving totem and phratry names through her, are of their father's local tribe. An alien woman may be assigned, by the elder men, to this or that totem: or to the totem corresponding to that which she had in her own local tribe. The children of male aliens follow the totem of their mother, a member of the tribe.

In Attica, too, was a local community, the deme—thus Thucydides was a Halimusian by deme. The historical demes were organised by Cleisthenes, on a local basis. Some of them bore the names of the γένη which occupied them, and often the names were derived from plants. Either these plants were characteristic of the localities, or conceivably the γένη had old totemic plant names, like the plum and other vegetable totems of the Australians. All about the local demes, the members of the phratriæ were scattered, like members of various totem names among the Australian local tribes. An alien could belong to a local deme, but not to a Φρατρία. His children, if by marriage with a free woman, were reckoned in her father's Φρατρία male descent prevailing, of course, in Attica. In

---

[266] J.A.I. xiv. 142, 181.
[267] Politics of Aristotle, Bolland and Lang, 1876; 'Family' in Encyclopædia Britannica.

Australia the tribes-woman's children by an alien would usually go to her totem and 'primary exogamous division.' The child of an alien woman, in Attica, even if the father was high born, could not be admitted to a Φρατρία: which certainly looks like a survival of the archaic reckoning by female descent. To try to insert an alien child in a deme was a civil, in a Φρατρία was a religious offence.[268] The ancient court of the Areopagus had to do with these offences against customary religion. Messrs. Fison and Howitt draw a parallel between the Areopagus and the Great Council of the Dieri tribe, whose headman was inspired by 'the great spirit Kuchi,' of whom one would like to know more.

An Attic boy was presented to his Φρατρία at once; full membership of the local deme came with adolescence, and after military training and service. As we know, a series of initiations, and instruction 'as to the existence of a great spirit,' with a probation of a year, are to be passed before the Australian lad is allowed to marry and attend the assembly of his local tribe. Better examples of initiation, and of a retreat in the hills in company with an adult, and instructor, are to be found in Sparta than in Athens. But the Australian 'and Attic analogies are pretty close. On the most important point there is no analogy. There were plenty of Φρατρίαι, of 'phratries' each Australian tribe has only two. Again, these two are exogamous: that is their main raison d'être. We have not a glimpse of exogamy in the Φρατρία of Attica.

The γένος, we may agree, I think, with Messrs. Fison and Howitt, was, originally, like the totem kin, an association of persons supposed to be related by ties of blood. The grammarian Pollux says 'they who belonged to the γένος were styled γεννῆται' (men of the γένος, and 'men of the same milk'), 'not that they were related γένει, but they were so called from their union (or assemblage—ἐκ δὲ τῆς συνόδον).' What is meant by γένει μὲν οὐ προσήκοντες 'not genealogically related'? I conceive Pollux to mean that the members of the γένος were not all of traceable or recognised degrees of kinship. Thus a Cameron, if asked whether he is related to another Cameron, may say, and not so long ago would have said, 'he is not my relation, but my clansman.' Messrs. Fison and Howitt take much the same view. By 'relations,' Pollux meant 'such as parents, sons, brothers, and those before them, and their progeny,' that is, from grandfathers and granduncles to grandsons and great-nephews. This might be the notion of relationship in the time of Pollux, the second century of our era, but, as Messrs. Fison and Howitt justly remark, Attic ideas of kinship before the συνοικισμὸς ascribed to

---

[268] Dem. Centra Neæram 17.

163

Theseus would be much more extensive, as in Scotland and Britanny. The humblest Stewart, Douglas, Ruthven, or Hamilton would call himself 'the King's poor cousin.' But the Greeks of our second century were more modern, more like the English.

Yet the very words γένος and gens indicate the idea of blood relationship, just as 'clan' does. The γένη had common sacra, and a common place of burial. They were clans, but we have no proof that they were ever exogamous or totemic. However, the myths and rituals of Greece certainly yield facts of which a totemic past seems the most plausible explanation. Mr. Jevons writes, 'we find fragments of the system' (Totemism), 'one here and another there, which, if only they had not been scattered, but had been found together, would have made a living whole. Thus we have families whose names indicate that they were originally totem clans, e.g. there were Cynadæ at Athens, as there was a Dog clan among the Mohicans; but we have no evidence to show that the dog was sacred to the Cynadæ.... On the other hand, storks were revered by the Thessalians, but there is nothing to show that there was a stork clan in Thessaly.'[269] Wolves were buried solemnly in Attica, where there was a wolf hero, and lobsters were buried in Seriphos, like the gazelle in Arabia. But we have no evidence of a wolf kin in Attica, though we have in Italy (the Hirpi) nor of a lobster kin in Seriphos. (For other traces, fairly numerous, I may refer to my Custom and Myth, and Myth, Ritual, and Religion, while deprecating the idea that all worship or reverence of animals is of totemistic origin.)

It will probably be admitted that, if Greeks (or ancient dwellers on Greek soil) were at some remote period totemistic, and next, by reckoning descent in the male line, became attached to localities, then something like demes, phratries, and γένη might very naturally be evolved. And many traces in ritual, myth, and custom do point to Totemism in the remote past. Indeed, it is remarkable that we should still be able to point to so many apparent relics of institutions already almost obliterated among the Melanesians.

On the whole, I regard it as more probable than not, that, in the education of mankind, Totemism has played a part everywhere; a beneficent part. But this is only a private opinion: one believes in it as one believes in telepathy, without asserting that the evidence is of constraining value.

---

[269] Introduction to the History of Religion, pp. 125-126.

PRIMAL LAW

CHAPTER I

MAN IN THE BRUTAL STAGE

Mr. Darwin on the primitive relations of the sexes.—Primitive man monogamous or polygamous.—His jealousy.—Expulsion of young males.— The author's inferences as to the evolution of Primal Law.—A customary Rule of Conduct evolved.—Traces surviving in savage life.—The customs of Avoidance.—Custom of Exogamy arose in the animal stage.—Brother and Sister Avoidance.—The author's own observation of this custom in New Caledonia.—Strangeness of such a custom among houseless nomads in Australia—Rapid decay under European influences.

'Man, as I have attempted to show, is certainly descended from some Apelike Creature. We may, indeed, conclude, from what we know of the jealousy of all Male Quadrupeds, armed as many of them are with special weapons for battling with their rivals, that promiscuous intercourse in a state of Nature is extremely improbable. Therefore, looking far enough back in the Stream of Time, and judging from the Social habits of Man as he now exists, the most probable view is that he aboriginally lived in small communities, each with a single wife, or, if powerful, with several, whom he jealously guarded against all other Men. Or he may not have been a social animal[270] and yet have lived with several wives like the Gorilla—for all the natives agree that but one adult male is seen in a band; when the young male grows up, a contest takes place for the mastery, and the strongest, by killing or driving out the others, establishes himself as head of the Community.

'Younger males, being thus expelled and wandering about, would, when at last successful in finding a partner, prevent too close interbreeding within the limits of the same family.'[271]

Mr. Darwin, in the foregoing sentences, affirms the improbability of Promiscuity in the Sexual Relations of Man during

---

[270] Mr. Atkinson's theory is based on the idea that our supposed anthropoid ancestor was eminently unsocial.—A. L.
[271] Darwin, Descent of Man, ii. 361-363 (1871).

the Animal Stage, and, incidentally, the Unity of the Human Race in its origin. Both theories are contested. The following thesis, however, on the Genesis of Primal Law in Human Marriage, treats of a conjectural series of events in the Ascent of Man, events which involve a state of the inter-sexual relationships amidst our primitive ancestors identical with that portrayed in the Descent of Man. My essay includes further, as regards the continued evolution of society, the development of a theory, based on my 'Primal Law,' which, if correct, would seem also to confirm Mr. Darwin's ideas as to Unity of Origin.

I am content, for my part, to hope that my hypothesis, however novel some of its conclusions, is in its general tenor in accord with the views of so great a naturalist as Mr. Darwin. His exposition of the probable relations, within the family group, of the male and female prototypes of mankind, and more especially of the antagonistic attitude, inter se, of the older and younger males, is indeed literally prophetic of the Primal Law, whose existence I surmise. This law is the inevitable corollary of Mr. Darwin's statement, if Man was ever to emerge from the Brute. My theory, in fact, viewed as to its genesis, is simply evolved from a consideration of the potential results of the attitude of such creatures as our ancestors then were, when subjected to the effects of those changes of environment, which alone, to my deeming, could have fixed modifications towards the human type. Mr. Darwin's premises, indeed, as to the Early Social economy of our Race in the animal stage, inevitably entail, if progress was to be made, the evolution of law in regulation of Marriage relationship, having regard to the fierce sexual jealousy of the males, on the one hand, and on the other to the patent truth that in the peaceful aggregation of our ancestors alone lay the germ of Society.

This would above all be the case if, reasoning by analogy, we provisionally accept, as the probable nearest approach to man's direct ancestors, the actual Anthropoids. These, such as the Gorilla, are undoubtedly amongst the most unsocial of animals as regards the attitude of the adult males inter se. From the very difficulty of the problem of the congregation of such creatures in friendly unison within the group, we may infer that, in its solution, there will be found the key to the whole question of the Ascent from Brute to Man. In that ascent, Habit, the parent of Law, must have been conquered, and modified into the direction of novel Custom, a shock to the older economy of life. Again, the new rule of conduct, necessarily inchoate (considering the presumed feeble intellectuality of the creatures concerned, animals more or less brutish) must yet be of facile interpretation to its subjects, though,

166

as befits Homo alalus, it must have been quite mute in operation. The new Rule of Conduct would not be expressed in terms of speech, a function, ex hypothesi not yet evolved. The rule, as it was to my mind, I here propose to attempt to unfold as the 'Primal Law;' hoping to show that therein lay the beginning of law and order, and that, whilst itself arising in a natural manner, in its incidental creation of a first standard of a possible right and wrong, it laid, so to speak, one of the foundations of that moral sense, which has seemed to place so wide a space between man and other creatures.

The prior existence of this law, in the semi-brutish stage of our physical and ethical evolution, might have been deductively evolved, even if no traces of it had remained to our day. It will be, however, my endeavour to point out that evidence of its existence (abundant as it appears to me) is to be found in certain obscure customs which are common to most actual savage races. The customs of so-called 'avoidance' between near relations will have the principal interest for us, although primitive marriage and inheritance will be found of corroborative value. Survivals and myths can be shown to point to the undeniable occurrence of this 'Primal Law' in the earlier life-history of the non-civilised peoples. The myths, however, may be merely early guesses about the unknown past of the race.

Amongst marriage customs that which has given rise to most discussion as regards its origin is 'Exogamy' or marriage outside the family group, or outside the limit of the totem name. My general argument, as will be seen, places me in antagonism with all theories yet advanced on the subject. But Mr. Lang, in Custom and Myth, 1884 (p. 258), hazards, as his own impression, a conception of this matter which I will note—namely, that 'Exogamy, may be connected with some early idea of which we have lost touch,' and he adds, 'If we only knew the origin of the prohibition to marry within the family[272] all would be plain sailing.' However utterly beyond human ken, in these our latter days, any truthful image of so remote a past may seem to be, it is yet precisely this hypothetic early idea which I hope to be able to expose. (If I am correct, we shall find that it was connected with the sexual relations of primitive man, whilst in the animal stage, and especially with the mutual marital rights of the males within a group.) Such idea in travail, hastened and sharpened by needs of environment, created issues which necessarily gave birth to a 'Primal Law' prohibitory of marriage between certain members of a family or local group, and thus, in natural sequence,

---

[272] I ought to have said 'within the community, whether local or of recognised kindred, indicated by the totem name.'—A. L.

led to forced connubial union beyond its circle the family, or local group—that is, led to Exogamy. But if such was in reality the original order of succession in the growth of custom, it becomes evident that Exogamy as a habit (not as an expressed law) must have been of primordial evolution. Thus (in contra-distinction to generally received opinion and to Mr. McLennan's theory in particular) Exogamy must have been a cause rather than an effect in relation to its ordinary concomitants, i.e. Female Infanticide as a custom, Polyandry as a fixed institution, and Totemism as connected with exogamous groups, within which marriage was forbidden. As thus my new hypothesis finds itself in opposition to those of recognised authorities, it is evident that it will require to account for all the facts if it is to hold its ground.

However convinced the author may be by the array of seemingly confirmatory details in favour of his hypothesis, it is possible that from their paucity they may yet to others seem to constitute but a feeble line of defensive proof. But if the theory shall prove in itself to have merit, this defect (arising, as I believe, from lack of general anthropological knowledge on my part, for I dwell 'far from books') will quickly be remedied, for a hundred other details in favour of my view will be at once perceived by more experienced students. Should my hypothesis really furnish the clue to the problem of the prohibition to marry within the family name, or totem name, all the rest will doubtless become 'plain sailing' in competent hands.

In any case before my conjecture is definitely laid aside as erroneous, it may, let us hope, be considered desirable to await fuller evidence as to the extent of the operation, in actual savage life, of that particular custom of 'Avoidance' which I consider, in its inception, and as the earliest law, to have been a 'vera causa' of widest operation in primitive social evolution. 'Avoidance' is, however, to-day, a mere faint image of a remote past, and its genetic significance has utterly faded from among even those people who yet, with strange conservatism, still blindly yield an everyday obedience to it, in form at least. Belonging to a class of savage habits presenting features so extraordinary, 'Avoidance between brother and sister' has ever been a puzzle to inquirers. This Avoidance is only the most obscure of all the numerous cases of the strange habit, but it is also that which, up to the present, seems least to have attracted the notice of anthropologists. In this class of custom, the Avoidance of which most frequent mention has been made in literature, is avoidance between mother-in-law and son-in-law, whereas that between brother and sister is to my knowledge but

168

rarely mentioned.[273] And yet, as far as my own experience goes (and it extends over more than a quarter of a century among primitive peoples in the South Seas), Avoidance of brother and sister is not only as common as, but infinitely more strict and severe in action than, the Avoidance of 'Mothers-in-law.' It is indeed probable that the very severity of observance has led to its being so little noticed. For by the action of this law, a brother and a sister, after childhood, are kept so far apart from one another, that only those who have actually lived long amidst natives can be expected to have had a chance of being aware of the restraints to intercourse between them. Even then it would be from some such casual occurrence as the accidental rencontre of the two, placing them thus in sudden and unavoidable proximity to each other, which would lead to an observation, by an European, of their extraordinary attitude and behaviour under such circumstances.

My own attention was primarily only drawn to this matter by noting the grave scandal and excitement caused in a native community by the momentary isolation, in a canoe, of a brother and sister. The affair became so very serious for the brother that he disappeared from the tribe for over a year. Indeed, the rigorous severity of this particular law in daily action is almost incredible. In New Caledonia, for instance, all intercourse between a brother and sister by speech or sign is absolutely prohibited from a very early age. Whilst the girl will remain in the paternal home, the boy, at the age of seven or eight (when not, as is usual, adopted by the maternal uncle), only comes there for his meals, partaken again solely with the other males.[274] He dwells until married in the large general bachelors' hut, set apart for youths in all villages. Even after marriage, if brother and sister have to communicate with each other on family matters, such communication must be made through the intermediary of a third person, nor can the sister enter the brother's hut even after his marriage, despite the presence of her sister-in-law therein. If the two should unexpectedly meet in some narrow path, the girl will throw herself face downwards into the nearest bush, whilst the boy will pass without turning his head, and as if unaware of her presence.

They cannot mention each other's names, and if the sister's name is mentioned publicly before the brother, he will show much

---

[273] This was written before the appearance of Mr. Crawley's Mystic Rose (1902).
[274] Cf. V. de Rochas, La Nouvelle-Calédonie, p. 239; Crawley, Mystic Rose, p. 217.

embarrassment, and if it is repeated he will retire precipitately. She can eat nothing he has carried or cooked. Whilst, then, such propinquity as is implied in the mutual habitation of the same hut by these two would be scandalously impossible, it is not uncommon to find a mother-in-law and son-in-law, whilst in Avoidance, living under the same roof. It is obvious that in the latter case each detail of 'Avoidance' in act or speech would be easily remarked by Europeans, whereas no chance of such observation between the adult brother and sister could possibly arise, they being kept, as we see, so utterly apart. It is to be noted, however, that the seemingly instinctive natural affection between two so nearly related is not quenched by these strange restraints. They remain interested in each other's welfare, and in cases of sickness, for instance, keep themselves informed of each other's condition through third persons. So great, however, is the depth in action (on these lines) of the feeling of avoidance in this matter, that I am convinced that the infanticide of twins, which only takes place in New Caledonia when the children are of different sexes, arises from the idea of a too close propinquity in the womb. Further evidence as to the very widespread existence of this custom in the South Seas I will leave to a later stage, only noting here that I have been astonished to find, in answer to inquiries, that it is well recognised amongst the aborigines, of Australia.

[Mr. Atkinson has left a blank space for an expected communication from the late Mr. Curr. On 'Avoidance' in Australia, between brother and sister, Messrs Spencer and Gillen write: 'A curious custom exists with regard to the mutual behaviour of elder and younger sisters and their brothers. A man may speak freely to his elder sisters in blood, but those who are tribal Ungaraitcha must only be spoken to at a considerable distance. To younger sisters, blood and tribal, he may not speak, or, at least, only at such a distance that the features are indistinguishable.... We cannot discover any explanation of this restriction in regard to the younger sister; it can hardly be supposed that it has anything to do with the dread of anything like incest, else why is there not as strong a restriction in the case of the elder sisters?[275]

Now the occurrence of this particular habit amidst a race of nomad hunters, forced by the exigencies of the chase to wander about in isolated groups, composed for the most part of single families, and where the separation of the sexes cannot possibly be arranged, as with the hut and village dwelling Caledonians, is a

---

[275] Native Tribes of Central Australia, pp. 88, 89.—A. L..

170

most remarkable fact. When we take into consideration the disturbing effects of such an avoidance in the internal economy of such a family circle, the significance of the circumstance is great as regards our general argument. It becomes, indeed, evident that the fundamental cause of the custom involving this daily and hourly dislocation of domestic life, must lie very deep in savage society. If, however, our theory as to the idea which dominated the inception of this strange habit shall turn out to be correct, then it will be seen that no surprise need be felt, if the genesis of this rule should prove to be in the animal stage, that traces of the superstructure should exist to our day. Now that attention will perhaps be more closely drawn to this, till recently the least observed of the cases of Avoidance, I feel sure that proof of its existence will be found in abundance in the present or past of all primitive peoples.[276] In view of its unexpectedly wide dissemination in Australia, hope may be felt that research will find it as a working factor in many peoples where its presence has been least expected, and not only in Australasia. It is possible that a stricter examination of the inner life of lower races in Africa and Asia will allow a perfectly legitimate inference that they are still under the influence of its effect, although the custom itself may be no longer in actual force. It is also possible, as I have said, that Survivals and Myths may point conclusively to its having had its day amongst the highest nations, with whom all traditions of it have been lost before the dawn of history. [Rather the reverse is the case; see the marriage of Zeus and Hera, brother and sister, and of the Incas, &c.—A. L.]

In many cases philological evidence based on the derivation of the root syllable of the word 'sister, a word which in the tongues of peoples still obedient to this law is from a root implying 'Avoidance,' may afford affirmative proof, as circumstantial as unexpected, that this custom was once as universal as my theory would require.'[277] If difficulty is felt in the acceptation of an hypothesis of such wide significance, simply based on an obscure lower custom so little noted in anthropological literature as to permit doubts of its existence, I can only repeat that a cognisance of the traits of this particular habit of avoidance and its effect as a factor in savage life demands such conditions of residence and chances of observation,

---

[276] Mr. Atkinson's forecast was correct. Brother and sister avoidance is very widely diffused.—A. L.

[277] The author does not give examples of words for 'sister' implying avoidance. But we elsewhere show that in Lifu (Melanesia) the word for 'sister' means 'not to be touched.'—A. L.

as can fall to the lot of few. I may add that it is one of the very first customs to disappear after contact with whites, especially missionaries, being, as it is, in such extreme divergence with the economy of the European family, in regard to the mutual attitude of brother and sister.

It is more than a quarter of a century since the author had his attention first drawn to the practice. The evolution of the idea of its possible identity with the Primal Law has led to a continued and close observation; he is thus able to certify as to its rapid disappearance. Brother and sister avoidance was at that time, thirty years ago, quite universal in New Caledonia; now in many places it is unknown, even as a tradition, among the younger aborigines. In view of the probability of a similar oblivion among other peoples, the immediate collection of evidence is urgent, and further delay seems dangerous and even culpable.

Thus, however much to the present advantage of the theory as regards the custom it would have been to cull larger proofs from that vast field of literature only to be procured in older lands, it has seemed desirable to make this thesis public without further delay. As we have said, if the theory is correct, wider students will bring forward cogent facts in further proof from existing knowledge, whilst continued research should afford evidence so complete of the widespread existence of the custom in the present and past of the human race, as to render my speculation as to its origin less seemingly illegitimate.[278]

CHAPTER II

SEXUAL RELATIONS OF ANIMALS

Brother and Sister Avoidance, a partial usage among the higher mammals.—Males' attitude to females in a group dominated by a single male head.—Band of exiled young males.—Their relations to the sire. —Examples in cattle and horses—In game-fowl.—Strict localisation of animals.—Exiled young males hover on the fringe of the parent group.—Parricide.

---

[278] Other speculations have now been advanced, especially by Mr. Crawley.—A. L.

Another difficulty in connection with the evolution of the so-called Primal Law of Avoidance between brother and sister from that early idea which we will presently disclose, seems to lie in the fact that if, as we uphold, such law was the first factor in the ascent of man, it must have taken its rise whilst he was still some ape-like creature. It remains, however, to be shown from its peculiar form that in its primitive application, the law would not have required for its intelligence greater mental power than is possessed by actual anthropoids. The law may indeed be said to be practically an inchoate fact, an actual if partial usage, for the regulation of the intersexual relations among most of the higher mammals. It could, at any rate, have come into full intelligent application as a well-defined social institution, in the actual sense of the term, whilst the anthropomorphic progenitor of man was still so little removed from the ape that

His speech was yet as halting as his gait,
Only less brutish than his moral state.

Briefly, the law of Avoidance concerns (more particularly) the relation 'inter se,' from a sexual point of view, of the male and female offspring of any given parents. In other words it determines the mutual attitude to the females within a (single) family group dominated by a male head.

Before, however, entering into the argument in this connection, it will be desirable to make a paraphrase on Mr. Darwin's dictum as to the social condition of man in the animal stage in general, and more particularly in regard to his intermarital relations, and to compare this with that of actual mammals. It is to be noted that he does not pronounce definitely as to whether, in the era of pure animalism, the original type of man's ancestor was social or non-social in habit. But we may judge from the extract already made from the Descent of Man that Darwin evidently inclines to the opinion that, even primitively, he was a social animal, as seemingly more in accord with the present eminently social conditions.

The very significant counter-fact, however, remains, that none of the actual anthropoids, as far as regards the adult males, are in any way social or even gregarious; the conclusion thus seems evident that, like these his nearest compeers of to-day, man was on the contrary a non-social animal, and that, as with the gorilla, only one grown male would have been seen in a band. We must then imagine our more or less human ancestor, roaming the forest in search of daily food, as a solitary polygamous male, with wife or

173

wives and female children; the unsocial head of a solitary isolated group.

With equal strength and probably already greater cunning than any actual animal of to-day, he had perhaps acquired dominance over most of the other beasts of the field. The patriarch had only one enemy whom he should dread, an enemy with each coming year more and more to be feared—deadly rivals of his very own flesh and blood, and the fruit of his loins—namely, that neighbouring group of young males exiled by sexual jealousy from his own and similar family groups—a youthful band of brothers living together in forced celibacy, or at most in polyandrous relation with some single female captive.[279] A horde as yet weak in their impubescence, they are, but they would, when strength was gained with time, inevitably wrench by combined attacks, renewed again and again, both wife and life from the paternal tyrant. But they themselves, after brief communistic enjoyment, would be segregated anew by the fierce fire of sexual jealousy, each survivor of the slaughter relapsing into lonely sovereignty, the head of the typical group with its characteristic feature of a single adult male member in antagonism with every other adult male. Now it can be shown that this vicious circle of the stream of social life is common to most mammals. The facts of the circumstance can be most easily observed amidst the half-wild, half-domesticated animals met with in colonial farming experience, in New Caledonia, for instance, where European horses and cattle have been allowed to return almost to a state of nature.

In this respect the economy of life in a herd of even such gregarious creatures as the bovine race, is a very curious and instructive study. There is no fact more striking than the subordination in which the younger bulls are kept; as long as they are at all tolerated in the herd by its patriarch, their intercourse with the females is most limited, and only takes place by stealth and at the risk of life and limb.

Nothing, as breeders are aware, is so fatal to the well-being of a herd, or leads so quickly to degeneration, as the perpetuation of the race by immature males. That procreation should be the act of the robust adult alone, is evidently an axiom with nature herself in successful production; it is doubtless of the highest importance to keep up the normal standard of strength and size. As a fact, the presence of the immature male among a herd of cattle is only permitted whilst he is still quite impubescent. Then banishment by the master of the herd is inevitable at a later stage. These exiles,

---

[279] Why 'single?—A. L.

although thus apart from the main herd, remain in touch with it, so to speak, and we find in consequence, in continual proximity of the troop of the patriarch and his females, a small band of males, which, as is evident from their colour and general physical resemblance, are its direct product. The relations between this mob and the old male are always strained, the latter has constantly to be on the watch to shield his marital rights.

For long the mere menace of his presence suffices for such protection, but with age—the young bulls becoming more bold—struggles take place which sooner or later, from mere force of numbers, end in the rout or death of the parent. We may here cite the mention made by Mr. Darwin[280] of Lord Tankerville's account of the battles of the wild bulls in Chillingham Park: 'In 1861 several contended for mastery, and it was observed that two of the younger bulls attacked in concert the old leader of the herd, overthrew and disabled him, so that he was believed by the keepers to be lying mortally wounded in a neighbouring wood. But a few days afterwards one of the young bulls approached the wood alone, and then the monarch of the chase, who had been lashing himself up for vengeance, came out and in a short time killed his antagonist. He then quietly joined the herd, and long held undisputed sway.' I may add from my own observation among half-wild herds in the colonies, that often when the old patriarch is not absolutely killed in such cases, he is forced to quit the herd. He then becomes a solitary exile, always exceptionally savage and dangerous to approach or molest. It seems to me probable that we have here an explanation of the occurrence of the existence of the well-known 'rogue' elephant, which is always a male and notoriously dangerous.

One important fact must here, however, be noted; that before such death or exile takes place, and the sons reach an age which enables them successfully to dispute the supremacy of the father, the daughters have reached puberty and borne produce to the sire—this matter, as will be seen later, has an important bearing on our general argument—on our theory of Primal Law. Amongst horses, again, which have become wild, exactly the same facts are to be observed. Each herd has one head, and this, as natural selection would imply, is the most powerful stallion; he is the master and owner of the females, and this mastery he retains until overpowered by other males, which, as before, are almost invariably his own progeny. In fact, any strange male would probably have first to run the gauntlet of this outlying herd of exiled sons before he could reach the father. It is, however, again to be noted that he is rarely

---

[280] Descent of Man, p. 501.

thus overpowered by even the combined efforts of his sons, before his daughters have reached such an age as to have produced offspring to their father.

This system of sequence of generations in breeding is, indeed, so universal in a state of nature amongst all animals, as to seem to point to the fact that in-breeding between father and daughter cannot be so prejudicial as some believe. Its efficacy in type-fixing is at least very great, if, as experiments of my own in pig-breeding on these lines would lead me to think, the question of prepotency is merely a matter of such close in-breeding repeated for generations. We may note here that if, as is probable, the produce, on the contrary, of a full brother and sister are degenerate, nature seems to attempt to prevent its occurrence. On the succession of the sons to the father's rights, speedy conflicts from sexual jealousy arise amidst the former and lead to a rapid segregation of the herd, in which the chances of own brother and sister continuing to mate are slight. Until this segregration, however, does take place, nothing is more curious to watch than the attitude and relations of these young males among themselves, the oldest and strongest claiming prior marital rights, but no more.

The same phenomena in social economy may be observed with even greater intensity in lower ranks of life than the quadrupeds. For instance, in a large flock of game fowl which I had an opportunity to observe closely for several years, during which they had nearly relapsed into a state of nature, there was an exact reproduction of all these details. There existed the same division into small family groups, each headed by an adult male, the same subordination imposed on the junior males of their banishment at puberty, as also their inevitable combined attack, when sufficiently powerful, on their paternal enemy. His death resulted in the same communistic assumption of his rights, with a subsequent disruption, from jealousy, into the typical smaller and separate groups.

We thus find an identical condition of the sexual relations between the females of a group and its older and younger male members to be common in the animal world—the domination, in fact, of an idea that might, in the person of the senior male, confers marital rights over the female members of the family, and an inchoate rule of action resulting therefrom; which bars from the enjoyment of such right each junior male. To hold that man, whilst in the animal stage, should form an exception to the general rule seems unreasonable. If, as we are inclined to believe, he was originally a quite non-social animal, the fact becomes more possible still that, as with modern types of anthropoid apes, each adult as head of a group was at feud with every other. As regards the social evolution, it would indeed seem most natural that, as Mr. Darwin conceived, the first step in progress should have been taken by animals already united gregariously, and thus already imbued with some social feelings. Strange to say, the path in advance which the ancestor of man, in the light of our hypothesis, was destined to follow, disclosed itself as an indirect consequence of the very intensity of his non-social characteristic. In fact, as I fancy mil become evident in the development of our argument, the only line of progress open to man was one inaccessible to animals of gregarious habits, judging by the economy of life in a troop of mandrills or baboons.

Having ventured to differ from the great naturalist on this point, I would with deference take exception to his further statement—'That the younger males, being thus expelled and wandering about, would, when at last successful in finding a partner, prevent too close inter-breeding within the limits of the same family.' This, if I understand it rightly, would convey the idea that this youthful band quitted the scene of their birth, and deserted entirely their original habitat in the forest. I cannot help considering it an error to imagine that such wanderings could thus be without bounds. Nothing seems to me more remarkable and irrevocable in savage Nature than the rigid localisation of all living things in her realm. No fish in the sea, no bird in the air, but has its local habitation, which only becomes free to the stranger on the death of the occupant.[281] No corner on earth but seems to hold its lawful tenant, and the bounds thereof are defined within rigid limits. Within, there is safety, with a sense of ownership; without, is the great unknown, possessed by others, fiercely ready to defend their rights, and threaten every danger and death to the stranger intruder—unless quite otherwise formidable than adolescent youth.

It is thus probable, in fact, that in common with the lower animals, the band of exiled young males of our anthropoid ancestor haunted the neighbourhood of the parent herd, remaining thus on familiar ground, and in hearing of friendly voices. For we must remember that their feud was only with the paternal parent. In the magic alembic of time the constantly increasing shadow of their presence would take sudden dreadful form, but in parricidal crime alone. The sequel in disastrous incest, which Mr. Darwin would here conjecture at, Nature alone has ever been impotent to deal with. The problem of an effectual bar to undesirable union between brother and sister was solved by man alone, and in the Primal Law. An

---

[281] This fact is well known to anglers for trout.—A. L.

effort of his embryonic intellect, thus early defiant of Nature, the law placed ethically, for once and for ever, a distinction between him and every other creature.

CHAPTER III

MAN VARYING FROM ANIMALS

Effect of the absence of a special pairing season on nascent man.—Consequent state of ceaseless war between sire and young males.—Man already more than an ape.—Results of his prolonged infancy and of maternal love.—A young male permitted to live in the parent group.—Conditions in which this novelty arose.

In common, then, with their nearest congeners of to-day, we have found each male head of a group of our anthropoid ancestors in direct antagonism with every other male, and a consequent disruption of the family at each encounter with a superior force. This disruption, in its effects on a species of non-gregarious habits, would result not only in the dispersal of its members, but in the destruction of what material progress in the accumulation of property might have accrued. As this would have included all germs of mechanical discovery, again doubtless due solely to the superior constructive faculties of the male, it is evident that advance in a race thus socially constituted was quite impossible.

Now this antagonism of male with male, with all its retrograde consequences, a struggle fierce enough in all animals, had a more intense effect on nascent man than on any other creature that had ever existed. An added force was caused by the disappearance in the nascent human species of that season of physical and mental repose, granted by Nature to the rest of creation, when not actually in the moment or season of rut. This ever-recurring but limited period, ordinarily appearing for a certain fixed epoch in each year, by the exigencies of supply and demand in the necessarily abundant food required for nursing mothers, had lost its date-fixing power with this new creation—Man. With the very first steps in progress would come his adaptation to a more or less omnivorous and consequently more regular diet. The consequent modification would be profound in the matter of sexual habit and appetite. Man needed no longer to put limits to the season of love and desire.[282] This was a crime against Nature, new in the history of the world, a crime which Nature would probably have avenged by race-deterioration or extermination, if the germs of mental power had not been already strong enough to lift him, Man, to be, of all creatures, almost completely beyond the influence of environment, thence of Nature herself.

The intensity of the evil led to its cure. In a state of society where literally every male creature's hand was against the other, and life one continual uproar from their contending strife; where not only was there no instant's truce in the war-fare, but each blow dealt was emphasised (fatally) by the intellectual finesse which now directed it, it became a question of forced advance in progress or straight retreat in annihilation as a species. However difficult it may be to imagine by what path such a creature was ever to emerge from the materialistic labyrinth in which we thus find him involved, it is sure that he neither could nor did remain there. A forward step was somehow taken, some road out of the maze was somehow found.

It remains for us to trace, by what dim light of custom and tradition we may, the faint trail of those momentous footprints, which, however lame and halting, took the strait and difficult way to a higher life. We may expect to find, as is but natural, that the path was one before untrodden. As man followed it, at first unconsciously, from the shoulders of this new pilgrim, predestined to worthier burdens, would fall some of the heavy load of the mere animal nature.

There was now, in fact, to be a break in the economy of animal nature, as regards that vicious circle, where we found an ever-recurring violent succession to the solitary paternal tyrant, by sons whose parricidal hands were so soon again clenched in fratricidal strife. In the dawn of peace between this father and son we shall find the signpost to the new highway.

Before going further, we may here state our assumption that, when our ancestor had arrived at this crisis in his life, a crisis involving the vast psychological step in advance implied in the development of society, and the intelligence necessary for the evolution of the law in its regulation, he was already somewhat more than ape. The animal stage as forming part of the ladder of ascent from brute to man would be marked by degrees of progression, each a step further removed from the original type. These very earliest steps we indeed propose to examine later in

---

[282] See Westermarck, History of Human Marriage, ch. ii., 1891. The subject is obscure.

detail, for the present we will suppose they have been taken, and that the influence of environment, under certain hypothetic conditions, to be also detailed hereafter, has fostered physical modifications towards the human type such as we found in the matter of rut. But in nature the relation is very close between the physical and the mental qualities. The advance in one would possibly lead to a corresponding development of the other. Each is the necessary complement of each. For instance, as Mr. Darwin has pointed out, while the lower extremities become more and more used for progression alone, so the upper, thus left free, would be specialised as prehensile organs, so becoming both valet and tutor to the nascent brain. To push our metaphor to an extreme, we may say that when Homo Alalus trod the new path, it was already as a biped in an upright attitude, thus leaving at least his hand free to point it out to others, for as yet his tongue, at least by the hypothesis, was inarticulate.

Our line of research as regards the new departure was at once narrowed when we indicated that it ended in the peaceful conjunction of father and son. Our path will lie in the examination of the question as to what possible series of natural circumstances, in the domestic life of the race, could lead to such conjunction, and what law in such an age could suffice for regulation of such association if formed. We shall have to examine more closely (as far as our imagination will aid us) the exact conditions of the family life of the semi-human group which we have supposed typical in that era, i.e. the small isolated band of anthropoids, composed of a single polygamous adult male with dependent wives and offspring. His possible relations with these, especially his attitude towards his male children, will interest us. Therein should certainly be found the desired series of circumstances entailing a critical situation, whose happy resolution shall furnish the clue to the problem of that possible aggregation on which all future progress depends. However strange it may appear, it will be found, as we have already said, that the abnormal conditions imposed by the unnatural modifications of the sexual functions have served as a means to the end of advance in progress. And as, by their action in the past, anthropoid man had become the most sexually jealous and intractable of all creatures, so it may be expected that the series of causes which shall have for effect the restraint of such excess of passion, will possess further vast potentiality of action. Such latency in potentiality is evidently indispensable when we consider that there is here concerned the evolution of law in opposition to nature, and its triumph for all time over the mere brute.

But first as regards the fact of the association of adult males on friendly terms within the group, which fact has seemed to us to constitute the whole problem of progress, it would on a hasty view appear as if it had been already found in the band of exiled sons which we have seen haunting the parent horde. Here we meet with that aggregation of individuals whose combination in peaceful union should apparently be the result of some law in regulation. This idea would even seem to gain support from the fact that all the members being brothers, and living most probably in a state of polyandry, we here appear to find fulfilled exactly those genetic conditions of primitive marriage imperative according to Mr. McLennan's theory of the origin of society. It will not, however, be difficult to prove that, at least at this stage in evolution, such a group would lack the most essential elements of stability. Their unity, in fact, as has been already pointed out, could only endure as long as the youthfulness of the members necessitated union for protection, and their immaturity prevented the full play of the sexual passion. The horde would inevitably dissolve under the influence of jealousy at the adult stage, especially if, as is probable, the number of their female captives had increased with the gain in years and power. The necessary Primal Law which alone could determine peace within a family circle by recognising a distinction between female and female (the indispensable antecedent to a definition of marital rights) could never have arisen in such a body. It follows that if such law was ever evoked, it must have been from within the only other assembly in existence, viz. that headed by the solitary polygamous patriarch, 'the Cyclopean family.'

We have said that this family would be composed of the male head and his wives, the latter consisting of captured females, and further, let us note, of his own adult female offspring, accompanied by a troop of infants of both sexes. The absence of male offspring beyond those of tender years would be another most notable phenomenon. These sons would, as we have seen, have been banished at puberty from the herd, in common with the habit of most animals.

Now we have surmised that at this stage our subject has been modified, both physically and mentally, to a certain extent in approach to the human type, and there is precisely one special modification which would have been of paramount importance in view of the problem of advance in progress. For if we may thus infer a certain increase in the longevity of the nascent race at even so early a stage in evolution, then that evidently entails a more prolonged infancy. It follows that, however precocious, the young males before exile must have passed at least nine to ten helpless years under their mother's care. But, again, the rise of superior intellectual faculties in general presupposes a decided increase in the powers of memory, and this agent, in connection with that of the longer companionship, would here set in movement, sooner or later, a psychological factor of strangely magnified force as compared with what it is in the mere brute—namely, human maternal love.

Separation, however caused, between this mother and her child would be far more severely felt than by any other animal. At the renewed banishment of each of her male progeny by the jealous patriarch, the mother's feelings and instinct would be increasingly lacerated and outraged. Her agonised efforts to retain at least her last and youngest would be even stronger than with her first-born. It is exceedingly important to observe that her chances of success in this case would be much greater. When this last and dearest son approached adolescence, it is not difficult to perceive that the patriarch must have reached an age when the fire of desire may have become somewhat dull; whilst, again, his harem, from the presence of numerous adult daughters, would be increased to an extent that might have overtaxed his once more active powers. Given some such rather exceptional situation, where a happy opportunity in superlative mother love wrestled with a for once satiated paternal appetite in desire, we may here discern a possible key of the sociological problem which occupies us, and which consisted in a conjunction within one group of two adult males.

We must conceive that, in the march of the centuries, on some fateful day, the bloody tragedy in the last act of the familiar drama was avoided, and the edict of exile or death left unpronounced. Pure maternal love triumphed over the demons of lust and jealousy. A mother succeeded in keeping by her side a male child, and thus, by a strange coincidence, that father and son, who, amongst all mammals, had been the most deadly of enemies, were now the first to join hands. So portentous an alliance might well bring the world to their feet. The family group would now present, for the first time, the till then unknown spectacle of the inclusion within a domestic circle, and amidst its component females, of an adolescent male youth. It must, however, be admitted that such an event, at such an epoch, demanded imperatively very exceptional qualities, both physiological and psychological, in the primitive agents. The new happy ending to that old-world drama which had run for so long through blood and tears, was an innovation requiring very unusually gifted actors. How many failures had doubtless taken place in its rehearsal during the centuries, with less able or happy interpreters! It is probable that, in the new experiment, success was rendered possible by the rise of new powers in nascent man. Some feeble germ of altruism may already have arisen to make its force felt as an important factor.

It is also certain that, with such prolonged infancy, there had been opportunity for the development of paternal philo-progenitiveness. It is evident that such long-continued presence of sons could but result in a certain mutual sympathy, however inevitable the eventual exile.

The love and care of a parent for his offspring is, after all, ethically speaking, the normal condition. Habitual desertion at too early an age would be fatal. Their dissociation, the abnormal and only one, took place under the influence of the strongest passion in nature, again largely exaggerated in primitive man. But in such an era purely physical characteristics would undoubtedly have also a vast influence in the development of the incident we have tried to depict. The fiercely solitary patriarch who first consented to the intrusive presence within his family circle of another adult male was, as I think we can prove, a being of abnormal physical power as compared with his fellows. For we have assumed satiety in desire to have been a powerful factor in the innovatory struggle we have witnessed. But such satiety implies extensive polygamy, and yet again a large harem composed exclusively of unwilling outside captives is incredible, escape for them in the primeval forest being too facile. Thus the harem would certainly be formed of the female offspring of the tyrant himself. These alone would need no watch or guard, for them the unknown outside world was hostile ground. But again very many adult daughters imply a father stricken in years. That one of such advanced age, in an epoch when force was all in all, could, defiant of rivals, still retain possession of his female kind, presupposes vast enduring physical power, or at least the protective tradition of past exceptional strength, still enduring in terror. If, again, at so early a date in the history of man we may be permitted to surmise any development of the faculty of psychogenesis, then we may again perceive how extreme physical qualities might have facilitated the solution of the problem of the admission of the intrusive male. For it is credible that long undisputed supremacy of power as the result of personal vigour might, in its incredulous contempt of a possible rivalry, show a tolerance of a situation utterly impossible to a weaker nature.

CHAPTER IV

EARLIEST EVOLUTION OF LAW

Trace between semi-human sire and son.—Consequent distinction taken between female and female, as such—Consequent rise of habit of Brother and Sister Avoidance.—Result, son seeks female mate from without Note by the editor.

In what, then, we are willing to concede, must have been exceptional circumstances, may thus have been taken that first step in progress which was to lead to such vast advance. In a development of the social qualities depended the whole future of mankind, and here we seem to see their germ and birth.

When, however, we affirm that the triumph of maternal love in the continued companionship of a male child, constituted the solution of the social problem before us, we do not intend to convey the idea that it lay solely in the fact of a simple inclusion of male offspring within a group. Such a condition, however significant in the actual case, has nothing in itself but what is common to the family economy of many animals. It is the normal one, for instance, among many pithecoids, as baboons, &c., where we find the younger males still form an integral part of the horde, although denied all marital rights. But, however inexorable among such species the temporary separation from the females during the actual season of rut, there is at other times a propinquity in amity as members of the same herd, which lessened doubtless the fierceness of the strife during the periodic play of passion, a truce in fact admitting of peace and alliance in offence and defence during most of the year.

With our ancestors there could be no such healing pause, the unnatural sexual modification of the race had rendered it impossible. The non-periodicity of the sexual function in rut would have made the whole year, with two adult males in presence, an interval of trial insupportable to the mere brute. With this race the banishment of the youth would be for all time, and the loss would be not only that of an ally, each exile would become an active enemy. Now we have hinted that the importance, in a potential sense, of a movement towards union, in such creatures, arose precisely from the fact that, on account of the intensity of the relations between male and male, and especially between father and son, their amicable conjunction was only possible under such exceptional conditions as would probably conduce to its stability whenever it did take place.

184

Indeed such inchoate rule or habit, a corollary of the early idea, as reigns in regulation of marital rights among lower creatures, would not be fully adequate for this higher creation. With lower creatures, might alone confers rights, which feebler force ever seeks each chance to invade, all stratagems being legitimate as a means to that end. With inchoate man such imperfect rule of action had become utterly impossible. The greater endowment in memory and reason entailed a too fatally added hate on non-compliance. For inchoate men the requisite law required such further exactness in definition as should leave no doubt of a meaning, not only to be understood, but to be accepted and obeyed unconditionally.

For between this father and son there was yet no real peace, only a truce, and that enduring but so long as the latter respected those marital rights of the former which we found extending over all that was feminine in the horde. The intelligent acceptance by the intruding junior of the sole right of the senior to union with the females of a group, was its sine qua non, which the dawn of intellectuality in the race as inevitably imposed as it happily permitted. Such a step in advance as a possible obedience, ex animo, to such a law would be immense. Therefrom would issue the vital point of a conception of moral reserves in marital rights as regards the other sex; the germ of a profound and fundamental difference between brute and man. For the first time in the history of the world we encounter the factor which is to be the leading power in future social metamorphosis, i.e. an explicit distinction between female and female, as such. The superlative fact, indeed, in relation to our general argument, appears—namely, that certain females are now to become sacred to certain males, and that both (nota bene) are members of the same family circle.

But what shall be, in such an age, the notes of a law conveying this noble sentiment of sanctity, which, disarming jealousy, could permit peace where before strife reigned? How give the outer expression of the inner feeling, now aroused, of a change in the past intersexual attitude of certain group members? Whence borrow the eloquence which shall ordain rules in restriction of intercourse whilst yet, for Homo Alalus, they must needs be mute in expression.[283] In the primal law alone, as I hope in its portrayal to show, can each condition be comprised and found as such. It will be marked and recognised by a physical trait whose presence is as significant and imperative as it is characteristic of the epoch. For a sentiment of restraint in feeling, whilst articulate speech was yet lacking, could only be expressed by restrictive checks in act and

---

[283] How do we know that homo was still alalus?—A. L.

deed, requiring mere visual perception for interpretation—acts we may here note, which, as insulating the individual, would also inevitably tend to consecration.

Now we mentioned in our first chapter that, in connection with the primal law, certain cases of so-called avoidance, and especially that between near relatives, would have interest for us and probably afford aid in proof. We drew attention to the strange features marking these customs, which had rendered their origin a source of wondering conjecture to all inquirers. It may be that precisely the actual anomalism of these characteristics may render them eloquent in our case. In view of our past argument, in very deed, nothing now becomes insignificant in these quaint rules of non-propinquity between certain near relations; nothing inexpressive in the ordinance of non-recognition between individuals well known to each other; nothing not suggestive in the dread of mere contact between those whom nature would place closest together, no lack of import in the strange taciturnity so incongruous with our garrulous later days of unloosed tongues. There is a possible vestige of a past era of dumb show in their eloquent muteness; of connection in their actual utter unreason with a long dead past of all unfamiliar habits and manners. Further, is verily aught lacking, in these latter-day customs of avoidance, of the necessarily archaic features of a possible primeval law? If these in truth were still existent, would they not, with such traits in common, be simply classed with those? Undoubtedly so, as it seems to me.

Now in the course of our argument it has appeared that the inclusion of the son as the second adult male in a group would evolve, as the most primitive rule of action, restriction of intercourse between its component females and the intruder. But in such a group, the former would necessarily be to the latter in the relation of mother and sisters. Such restriction, again, taking the only possible form, would be avoidance of these relations, and thus there is a concurrence in resemblance with that particular habit of avoidance on which we enlarged in our first pages, viz. that between brother and sister (and now less strictly), between mother and son. Do we not thus seem to lay a finger on an actual law, still an every-day working factor in savage life, which is not only identical with, but is in very deed the primal law itself, in form at least? The acceptation of such intolerably irksome restraints as avoidance, in the daily economy of savage life, has seemed forcibly to imply a fundamental cause of profound depth. This cause now seems laid open to us. The unaccountable and seemingly unreasonable restrictions on intercourse which mark it thus betray their appropriate origin in a time of comparative unreason.

This then, the primal law—avoidance between a brother and sister—with appalling conservatism has descended through the ages (in conservance of form, if not of ultimate purpose). It ordained in the dawn of time a barrier between mother and son, and brother and sister, and that ordinance is still binding on all mankind [but in Egypt and Peru, for example, the opposite of this rule, for special reasons, has prevailed]. Between these for ever, a bit was placed in the mouth of desire, and chains on the feet of lust. Their mutual relationship is one that has been held sacred from a sexual point of view, in most later ages. It only remains for us to repeat that it follows that this law, as applied in the group composed of a single family, is, as we pointed out, the parent of exogamy; continuance within the group necessarily and logically entailed marriage without; but, again as we said, it was itself the offspring of the early idea. For this idea, in its assumption that sovereignty in marital right was compatible with solitude alone, was shaken to its depths when a second presence threatened rivalry, and demanded remedy in the action of law, which it has seemed to us could only take the form we have tried to portray[284] in the primal law.

NOTE TO CHAPTER IV

To the Editor this theory seems worthy of the ingenuity of his old friend and kinsman. Granting that early man was a speechless jealous brute, dwelling in groups consisting of a patriarchal beast, and all the females whom he could catch, and all the females whom he could beget; granting that he drove all his adolescent sons out; and finally (under the circumstances described) kept one or a few sons at home, his rule would tabu all females of the group to these sons. Otherwise there would be a fight.

The sons would have to bring in mates from without—the result is Exogamy. But Mr. Atkinson does not observe the numerous tabus existing among savages, on ordinary (not sexual) intercourse between men and women; as if each individual, of each sex, was or

---

[284] Later, as we further analyse the chords in the great hymn of human existence, we shall find that this first of all rules of intelligent moral action, however little it may have had of ethical intention in its inception, will ever remain (in its effects) the fundamental note in the harmony of psychical life. All succeeding law is its inevitable corollary, and vibrating in cadence with this fundamental note.

might be dangerous to each individual of the other sex; that is no idea of our speechless brute ancestors, of Mr. Atkinson's hypothesis. These tabus do not amount to absolute avoidance, but they do amount to very marked restrictions; for example, on eating together, or sleeping under the same roof, even where husbands and wives are concerned. For the facts, to save repetition, it is enough to refer to Mr. Crawley's book, The Mystic Rose. Now if these less rigid tabus between the sexes (which Mr. Atkinson noted in his observations on the life of New Caledonian natives) arose in the general savage superstitious dread of everything not a man's or woman's own self, they might become more rigid as propinquity increased. The most dangerous female would be the female who had most chance of being dangerous, by virtue of propinquity, namely the sister. She would therefore be the most strictly barred. The closest of all relations, that of lover and lover, and man and wife, would be most severely guarded, as most dangerous, by tabus. All this would happen (granting the verifiable condition of savage superstitious dread) even if Mr. Atkinson's theory of our speechless beast ancestors' way of life were wrong.

We should probably find the effects (Avoidance and Exogamy) even if the primeval causes postulated by him never actually existed. Moreover any avoidance between mother and son that may exist (as in the case of the mothers of chiefs, in New Caledonia, and their sons) is perhaps no more than part of the general rule of restricted familiarity between the sexes, whether that rule arises from a superstition, or from the circumstance that men and women sometimes 'disturb each other damnably,' as Lord Byron remarked to his wife. It might be argued that the exogamous prohibition is only one aspect of the general totem tabu; and that, in the case of brothers and sisters, incest against the totem tabu needed to be guarded against (as most likely to occur) by precautions of avoidance peculiarly stringent. These precautions, then, would not necessarily come down from the time of our hypothetical speechless beast ancestors. They might come down from that time, but the descent, it may be objected, is not necessary. The rules might have arisen among men as human as we are—Totemists. On the other hand, it might be argued for Mr. Atkinson that his hypothetical groups must be infinitely older than Totemism. When totem names were imposed on the earlier groups, the totem name and mark would only be a method of distinguishing group from group, probably becoming the germ of later superstitions by which everything connected with the totem was tabued, in each case, to the groups bearing its name. Either alternative hypothesis is easily conceivable: on the whole, I prefer the theory that exogamy arose, or an exogamous tendency arose, as in Mr. Atkinson's hypothesis, and was later sanctioned by the totem superstition.

As to brother and sister avoidance, if there is an 'instinct,' as Westermarck thinks, against marriage between near relations, if 'close living together inspires an aversion to intermarriage' (pp. 352, 545), then the avoidance of brother and sister would make them especially apt to fall in love together. But they don't. Brother and sister, under the tabu, are the greatest possible strangers to each other. They have not 'lived in long-continued intimate relationship from a period of life at which the action of desire is out of the question' (Westermarck, p. 353). They have done precisely the reverse. So why they do not fall in love with each other is what we have still to explain. All the rigid systems of brother and sister avoidance exist, it would seem, to prevent what never would have occurred, had the young people been allowed to grow up together. For in that case they could have had (we are to fancy) no erotic desires towards each other; that is Dr. Westermarck's idea. But could they not? He tells us that, among the Annamese, 'no girl who is twelve years old and has a brother is a virgin' (p. 292). And the Hottentots do not 'marry out of their own kraals' (p. 347).

Then where are we, exactly? If there is 'a real powerful instinct' against love between persons who 'have lived in a long-continued intimate relationship' from childhood—why does the instinct fail to affect Annamese and Hottentots, for instance? And if to be absolute strangers to each other is apt to make two young people fall in love, why do New-Caledonian brothers and sisters never do it? (Compare Mr. Crawley, The Mystic Rose, pp. 444-446.)

CHAPTER V

AVOIDANCES

Results in strengthening the groups which admit several adult males.—Disappearance of hostile band of exiled young males.—Relations of sire and female mates of young males now within the group.—Father-in-law and daughter-in-law avoidance.—Rights as between two generations.—Elder brother and younger brother's wife avoidances.—Note on Hostile Capture.

If we can admit the argument as to the sequence of incidents which thus led to the primary amicable conjunction of two males within the same group, it is not necessary to enter very minutely into the exact manner in which it would grow into a habit and spread throughout the land. We may surmise that, in spite of the advantages presented, its progress would, from the isolation of the groups, and their mutual hostility, be very slow. This would specially be the case if, as with physical variations, this point of departure in social development was a purely individual one, and so had to spread from a single centre by natural selection acting through a beneficial variation. It is, in fact, difficult to conceive, in view of the series of the abnormal factors we have supposed necessary for the genesis of such evolution, that any coincident departure of the same nature would be likely to occur in any other centre. It is even certain that the full possible benefit of the innovation would not be able to make itself felt even in the group of its creators. It is easy to understand that, in spite of the shield-like love of the mother, there would be friction between father and son in such unfamiliar circumstances, not only novel to the individual, but unhabitual to the race. In fact, it may be taken for granted that on the part of the father there was at first only a sulky tolerance of the new arrangement, a tacit but very unwilling acquiescence in the presence of the son.

On the part of the son would exist a watchful reserve, with an ever-haunting sense of insecurity, born of a novel and precarious situation. Even on such terms, however, and with what little might be of conciliation between the two, it is evident that a momentous forward step has been taken: the powers of the group in offence and defence, as against outsiders, would be enormously increased;[285] the fire of youth and the wisdom of old age for the first time joined forces, and paternal experience comes to the aid of filial courage and ardour. On the death of the patriarch the family found a natural protector, and what potential germs of advance, material or spiritual, had been evolved, would remain intact.

The real significance of the circumstance of such conjunction will, however, be found to lie in the character of its consequences as entailing further progress. Thus we have suggested that the original innovation consisted in the toleration of the presence of a single male offspring. But the way was evidently thus paved for the acceptance, at least in later generations, of others of the young males, although at first only of those who, not too much rivalling the

---

[285] It is clear that, for this reason, natural selection would favour the new kind of group. The arrangement would be imitated.—A. L.

fathers in power, would offer least grounds for jealousy. Now if we may accept it as an axiom in the matter of social progress in this race, that everything depended on aggregation of numbers in peaceful union, then such renewed inclusion presents itself in an important light. When it grew into a habit, the vast increase in power with every succeeding generation to a group, which is implied in the fact of each male child counting as an unit of strength, becomes evident. The new superiority to the original Cyclopean form of family, with its solitary male head, is enormous. The extinction of the latter type would only be a matter of time it would finally result from the easy capture, by better organised rivals, of their females. With the gradual disappearance of those who clung to the old order, the leaven of progress would spread in permanence through the whole mass. It would eventually become the rule that all the male offspring should remain within a group, to form henceforth an integral part of it.

This result would be very important from another point of view. Such retention of sons would lead to the elimination of one of the greatest past elements of disorder—that band of exiled young males, which we found as a constantly menacing adjunct of the Cyclopean family, would cease to exist. But, again, a very slight reflection will enable us to perceive that such a modification as the presence of these celibate young males in the family circle must soon have entailed consequences in social evolution of a new and strange complexion, thoroughly embarrassing, in such an era, to those interested. Primitive social economy was now, in fact, to enter on phases presenting such possibilities of complication and disruption as must forcibly have led to the continued evolution of law in regulation. Such complications will become at once apparent on an examination of the probable sequence of events in the family life of the race. Such law in regulation will be shown to have been evolved, and, as before, to be still existent as a rule of action in these latter days, and with all those weird characteristics of mutism and general anomalism which prove its archaic origin.

Granted a group consisting of a patriarch whose marital rights extend over all its females, and of young males whose attitude,[286] from a sexual point of view, is marked by the strict reserve ordained by the primal law, it by no means follows that such celibacy of the young males would extend beyond the feminine element of their own troop. On the contrary, the whole of the outside world remains free for them to choose from. In fact, it is evident that it is there, in the world outside of the group, that their future mates must be

---

[286] With portentous endurance of custom towards these.

found. On the component females of the parent horde a ban has been for ever laid, but all else of womankind are free of the interdict; they are beyond the law, and 'Sin is not imputed where there is no law.' Here, then, in the outer world, would their wives be sought. The complication we have mentioned would arise when, after successful captures of females by the young males, captures which it is hardly necessary to state would have been 'hostile,' the introduction of their captives within the parental group took place. The presence of females not to be his own within a circle where all that was feminine had ever been his in undisputed right, would certainly stir to its depths the soul of the Cyclopean type of parent. Such a situation must in its inception have caused a friction full of menace to the new order of things.

The only solution would be, as we have said, in the further evolution of law in remedy. We shall, as before, expect to find the law ordaining restrictions on intercourse between certain individuals, and marked with the archaic characteristic of mere visual action being sufficient for its interpretation. Such, then, as it was, we still find it, in the habit still common with many races of avoidance between father-in-law and daughter-in-law. In mute avoidance between these two could peace alone endure in the new crisis. The new rule implied the development of the same respect by the father for the marital rights of the son, as we have seen the primal law to have had for effect as regards the paternal prerogatives. Natural selection would come into play in the consolidation of this new stage in legislative evolution. For the group which first adopted such a modus vivendi would gain so great an advantage with each generation, in point of numbers alone, as would quickly give it supremacy. On the other hand, the forcible infringement by the father on the rights of possession by the sons in their captives would simply result in the withdrawal of the sons and their women. Hence disruption of the group and a fatal retrogression to the archaic type with all the weakness implied in a sole male component.

Here then we find renewed, in act of custom, another bar to intercourse between certain individuals of different sexes. And not only as a peace-conferring covenant would the fresh step in progress be important. It marks another stride in advance from brute to man, in the further recognition of points of difference between one female and another from a sexual point of view, the genetic evolution of which sentiment, in the primal law, foreshadowed such latent potentiality as already distinctive of mankind alone. Social advance to this stage has entailed the genesis of law in definition of respective marital rights as between the two generations, viz. fathers and sons, but further evolution in regulation of the individual right, as within the generation itself, is evidently indicated. For all members of the latter, as is the case to-day with many lower people, would be considered, de facto, a class, in which all are regarded as brothers, own or tribal, whose interest in all things regarding their classificatory rights would be in common.[287]

Such would be more especially the case in respect to female captives, whose capture would be the act of all. Here sexual jealousy, if uncontrolled, would inevitably lead to repetitions of that violent segregation of the members which occurred under the same circumstances amidst their primitive prototypes—i.e. that band of isolated young males, contemporaries of and exiles from some Cyclopean family. We may, however, surmise that, now or soon, the general development of intelligence and advance in social feeling would permit the action of the necessary rule in remedy. That rule would doubtless take the form we still find existing to-day for regulation in parallel circumstances, a rule which simply accords priority of right in accordance with seniority in birth. Such right would in itself accrue naturally as with other animals, from the fact that superior strength is found with greater age. This prior possession is not incompatible with an amicable recognition of the privilege of later participation by others. If such recognition took place in favour of the rights of the juniors, whilst they again peacefully accepted the larger pretensions of the seniors within their class, then natural selection would again act in their favour by the elimination of groups unable to abide such conditions. The arrogation of sole possession could but lead to the disintegration of the troop.[288]

Another solution of the problem of rights as between brothers may here be noted: it is that which is common to such widely separated spots as New Caledonia and Orissa, viz. the law of avoidance between an elder brother and a younger brother's wife. It is one of the most strict and severe. It is, however, incompatible

---

[287] Herr Cunow, as we showed, regards the 'classes' (not the 'phratries') of Australian tribes as based on a rough and ready calculation of non-intermarrying generations.—A. L.

[288] See also Westermarck, pp. 458, 459, on the Khyoungtha, a Chittagong hill tribe. After marriage a younger brother is allowed to touch the hand, to speak and laugh with his elder brother's wife, but it is thought improper for the elder brother even to look at the wife of his younger brother. This is a custom more or less among all hill tribes, it is found carried to even a preposterous extent among the Santals.

with group marriage, which we are now dealing with.[289] It marks the genetic stage of monandry.

So far, then, we have thus traced the evolutionary process of group formation—and we seem to find confirmed that affirmation as to the primordial order of succession in the genetic growth of custom which I ventured to submit in my first pages, viz. primo: the existence of an early idea of concupiscent lust, distinctive of the male head of a group, which led to his pretensions in marital right over all its component females in necessary incestuous union; secundo, the evolution of the primal law (with what little of originally ethical intention is now immaterial), in protection of such right when threatened by intruders; tertio, its acceptance by the latter, and, as an inevitable sequel, their indispensable capture of outside females as sole possible mates.[290]

But then this question of the absolute necessity of the rape of strange females as mates by the young males of a group, opens up to view another remarkable coincidence of effect in custom, still enduring to our day. As such it may furnish a clue to a feature in savage habits, to which we have already alluded as the cause of more discussion, concerning its origin, than any other. For habitual hostile capture of females outside a group by its male members, with a coincidental bar to sexual union with its component females, seems simply a definition of that habit among many actual peoples which has been called Exogamy by Mr. J. F. McLennan. Hence comes the evident corollary to the argument that the primal law and exogamy stand to each other in the mutual relation of cause and effect. We stated that if this was in reality the case, and if here we have the origin of marriage outside the group, then the novelty of the view, and the fact that it finds itself in opposition to other theories on the matter, weighty from the eminence of their propounders, would still require the production of a clear series of proofs in its favour if it was to be accepted. Such proofs, however, we predicted, would with research be found abundantly. We hope that already in our thesis, as far as it has gone, we may be considered to have advanced some such testimony in the seemingly necessary identity of custom, in form at least, in a hypothetic ancient and an actual modern era. There is surely here more than mere fortuitous coincidence in social evolution.

---

[289] As a fact the 'classes' (probably distinctions, originally, of generations) do not, I think, indicate 'group marriage.'—A. L.

[290] Westermarck, ut supra, pp. 387-389, 546, agrees. For the opposite view, cf. Crawley, p. 367. Westermarck does not seem very sure of his own mind.—A. L.

It seems, indeed, a legitimate inference that the divers habits of avoidance which we have cited, intelligible only by their congruency with such phases of genetic growth of custom as we have surmised, whilst presenting features utterly anomalous as latter-day creations, are in reality of the archaic origin we would assign to them. Their extraordinary vitality, which becomes almost bewildering to contemplate, may be explained by the fact that, as the first steps in progress, they would be necessarily woven into the whole social fabric.

It remains to be seen if, in further unravelling its tangled web, other threads of actual custom may not be found as apparently eloquent of a far distant, unfamiliar past, in their present abnormal features; other usages in every-day lower (savage) life, which in the light of a primal law shall furnish an unexpected solution of many perplexing problems in social evolution. If it can be shown that their inception would have been in happy accordance with the resolution of necessary incidents in evolutionary progress, may we not legitimately infer both that such customs thus had their origin, and again that these incidents really occurred? Our further research into the development of social institutions will point out indisputably, that primitive society was now on the eve of a succession of events in social order, presenting quite a series of menacing complications—their resolution will seemingly entail inevitably the continuous evolution of law in remedy, which law would have presented features identical with the actual laws of avoidance and others.

NOTE TO CHAPTER V

Marriage by Hostile Capture

Mr. Atkinson accepts, for the excessively early stage of semi-human society with which his hypothesis deals, the necessity of procuring mates for the young bucks by capture from a hostile group. Now Dr. Westermarck writes, 'Mr. McLennan thinks that marriage by capture arose from the rule of exogamy;' and Mr. Atkinson holds that it arose from the necessity of the case. The old patriarch allowed no female born within his group to be united to his sons. Dr. Westermarck says, 'It seems to me extremely probable that the practice of capturing women for wives is due chiefly to the aversion to close intermarrying ... together with the difficulty a savage man has in procuring a wife in a friendly manner, without giving compensation for the loss he inflicts on her father'

(Westermarck, 368-369). He admits a period when 'the idea of barter had hardly occurred to man's mind,' But Mr. Atkinson is thinking of a state of affairs in which the idea of barter had not occurred at all. Even at Dr. Westermarck's stage of the dawn of barter, 'marriage by capture must have been very common.' But Mr. Crawley argues that because, in his opinion, 'types of formal and connubial capture' are not survivals from actual capture, therefore 'the theory that mankind ... ever, in normal circumstances, were accustomed to obtain their wives by capture from other tribes, may be regarded as exploded' (Mystic Rose, p. 367). This dictum does not affect Mr. Atkinson's theory. Semi-human beings, in the conditions imagined by him, might be obliged to get their wives by capture, whether existing types of so-called formal capture are survivals of actual hostile capture or not. If Mr. Atkinson accepts the formal abductions as survivals of real captures and so as proofs of his argument, and if such formal abductions are not survivals of real capture—still, as Dr. Westermarck says, even after the supposed stage of semi-human life, 'marriages by capture must have been very common'—in Mr. Atkinson's hypothetical still earlier stage, they must have been universal.

CHAPTER VI

FROM THE GROUP TO THE TRIBE

Resemblance of semi-brutal group, at this stage, to actual savage tribe.—Resemblance merely superficial.—In this hypothetical semi-brutal group paternal incest survives.—Causes of its decline and extinction.—The Sire's widows in the group.—Arrival of outside suitors for them.—Brothers of wives of the group.—New comers barred from marital rights over their daughters.—Jealousy of their wives intervenes.—Value of sisters to be bartered for sisters of another group discovered.—Consequent resistance to incest of group sire.—Natural selection favours groups where resistance is successful.—Cousinage recognised in practice.—Intermarrying sets of cousins become phratries.—Exceptional cases of permitted incest in chiefs and kings.—No known trace of avoidance between father and daughter.—Progress had rendered such law superfluous.

A superficial view of the group we have examined might, from its general resemblance in custom to others among actual lower types of man, lead to a hasty conception of perfect identity, from a social point of view, in nearly all other respects. We see that exogamy, hostile capture, group marriage,[291] and obedience to certain accepted rules of avoidance, are common to both, to the hypothetical semi-bestial and to the actual savage groups.

The impression, however, would be very erroneous.

In the former, the hypothetical archaic stage, still lurked as a festering canker, an archaic element in marital prerogative, which marks it as of an epoch in the life-history of our race when the brute still triumphed over the man, an epoch far removed from our own. It possessed a feature in connubial relations as between certain group members which placed a profound gulf between it and any existing form of these days—a trait which, whilst it endured, would tend to render all further social progress difficult, if not impossible. It barred the road to that next great gradation in sociological evolution which is implied in the friendly conjunction of groups in a tribe. The latter stage was a vast upward step, but still it was only one round in the ladder of ascent to man, and indeed derived its chief importance from this fact as such. The tribe was the real goal; there, only, could be found the vital quality of social stability to be conferred by peaceful connubium between united groups as opposed to hostile capture between isolated families. Each group must come to be in itself complete, and yet each must form the necessary complementary parts of the actual Tribe common to all lower races, with its typical divisional inter-marriageable group classes ['phratries']. The fatal bar to a higher platform was a heritage from the anthropoid ancestor, and, as such, eminently characteristic of an animal stage.

This odious inheritance was the habit of incest between father and daughter, which we have found to be common to all the mammalia as a dominant domestic feature. As a factor in evolution we have seen that it actually had as direct outcome the primal law itself, and thus, with a strange irony, it may be said to have so laid the foundation of an ultimate moral sense. In such or other action in the past it had, however, served its useful purpose. Its operation in the future could be but detrimental; so opposed to all advance does it become, that, as we shall find, it is to be finally swept aside so completely as to permit to some students doubts of its existence, though 'In Saturn's time such mixture was not held a crime.' Leaving no traces of action in actual usage save in such exceptions

---

[291] As to group marriage the editor cannot follow Mr. Atkinson.

as prove the rule, it will not be difficult to show that, in giving birth to the primal law, it doomed its own existence, and this apart from any ethical connection. The continued progress of society led almost mechanically to developments eminently inimical to its continuance as a custom, whilst again it would be found even injurious to the order of things as constituted in the earliest group-plus-tribe stage. If we bear in mind the axiom that, other things being equal, the largest assemblage of individuals in amity would have the greatest chance of survival, as possessing more numerous units of strength, then father-and-daughter marriage would be pernicious by preventing an assembly from profiting to a full extent by the productive powers of all its members. For such incest implies sole marital rights by a senior generation of males over a junior generation of females. As the latter would always, from mere disparity of age, outlive the former, it follows that, on the death of a father the daughters would remain unproductive, the only other males in the family group being their own brothers, and as such barred to them by the primal law.

This situation, in itself an element of weakness, became doubly so, if, as is probable, these young widowed females seceded to other and hostile groups with whom union to them was free. Such groups would in consequence be by so much strengthened, at the expense of their original circle. If, on the other hand, these widows remained in their own circle, their presence as useless mouths would be embarrassing, and a possible source of danger as a temptation to outside suitors. Again, celibacy being quite an anomaly in such an era, complications might arise from possible infractions of the primal law itself within a group.

But it is in special relation to the further movement in advance implied in the friendly aggregation of groups into a tribe that the effects of paternal incest would be most fatally felt. For while it reigned as a custom and a father usurped sole marital right over the whole feminine element, the immigration into the group of outside suitors for their hands would be impossible, their possession by the latter would be only possible after capture, which, being hostile, would tend to keep asunder the different groups. And yet in the next and higher stage of social evolution, as presented in the amalgamation of groups into a tribe, the acceptation of these outside mates in peaceful connubium is precisely the most characteristic feature. In later days they will be found as the male members of a certain 'class' in one 'phratry,' and, de facto, eligible in group marriage with all and certain females of the corresponding category as regards birth in another phratry, within an all-embracing tribe. As indeed with actual Australians, where, by right of birth alone each 'class' contains the natural born husbands of the wives of another 'class.' Such connubium is evidently impossible while incest flourished as a custom, it could only arise after its decay.[292]

It thus becomes necessary to study by what possible conjunction of affairs so desirable a result was arrived at. We will find that, however fortuitous the event of the primary inclusion of an outside possible suitor within a group, however timid and hesitating his entry, his presence there would be the signal of the beginning of the end. Now it is evidently hopeless to look for any voluntary acceptance of his claims by the living father, to whom the temptation to so easy a procuring of an inmate of his harem as his own daughter would be irresistible. There would be also on his side all habit and tradition, and with no direct group interest in opposition, the brothers being unconcerned. The initiative in change must then arise irrespective of him, and without the obstacle of his presence. This could only be possible thus after his death.

Now it is important to observe that precisely the embarrassment we have seen arise after this event must be a means to the end of the conjunction we seek. We have noted the danger of the situation under such circumstances; ineligible in union by the primal law with the remaining male element, which is composed of their own brothers, temptations to its infraction would be as frequent as fatal, on the part of the early widowed sisters. On the other hand, the anomaly of a celibate existence in the animal stage would tend to the secession of widows, so to speak, to hostile hordes, or to constant attempts at hostile capture by the outside

---

[292] I have here slightly altered Mr. Atkinson's terminology. As the passage stands in his manuscript he confuses totem kins with the Australian intermarrying 'classes.' In his manuscript the passage runs thus: 'In later days they' (the outside mates) 'will be found as the male members of a certain class generation in one group' (by which he means a 'class,' say Ippai, in a 'phratry,' say Dilbi) 'and, de facto, eligible in group marriage with all and certain females of the same category as regards birth in another group.' Here he obviously should have written 'eligible in marriage with all females of the corresponding category in the other "phratry" of an all-embracing tribe.' 'As indeed with actual Australians where, by right of birth alone, each totem group contains the natural born husbands and wives of another totem group.' This is not the case: men of one totem kin are not compelled to take wives from one other totem kin; but men of one 'class' must take wives of one other 'class,' and men of one 'phratry' must take wives out of the other 'phratry.'. To avoid confusion I have, in the text, inserted the correct terminology.—A. L.

suitor. But with the friendly entry of the latter and his acceptance as a group member, all these disturbing influences would at once cease; further, the value of an extra unit of strength in his presence would soon make itself felt.

Let us then imagine a band of brothers willing to aid in the sustenance of their widowed sisters, strong enough to defy their capture by others, and determined to frustrate any attempt at escape on their part. The inevitable result would be the attraction within their own circle of suitors for their hands. Now it is worthy of note that the feasibility of the process of such attraction and inclusion becomes more obvious when we reflect that if, as is probable, they belonged to a neighbouring group, they would thus by no means find themselves quite strangers in their new home. For it is precisely from near neighbours that their wives would have been captured by the males of the assembly they have now joined. These wives, in fact, would be probably own sisters to the immigrants. As such, then, we can understand an easier tolerance of their presence by the resident males, their new brothers-in-law; as brothers and sisters the primal law created such a bar in division between their own wives and the new comers, as put aside any possible chance of friction in jealousy.

Now the significance of the entry of outside males would be vast, from many points of view. In a general sense we here find that further aggregation of numbers in unison which we considered important, as prophetic of the present social condition of to-day. Again, there arises a renewed distinction of that difference as between one female and another, so peculiar to mankind. For here we see that for the first time a sister no longer ranks in exactly the same line, from a marital point of view, as a mother. A daughter, in fact, may now evidently have as mate other than the husband-father. As the primitive mind habituated itself to this idea, the first serious blow was dealt at the old parental prerogative. Again, in other ways, in other minds than own brother and sister, will this change in the old order of things be thus brought home—to no one more clearly than to the outside suitor himself, when, later, he becomes a father; the trains of circumstances leading to it are very curious, but would arise in a perfectly natural manner. The result in this connection would make itself felt by him in the next generation, with the advent to the adult stage of his own female offspring.

Is it credible, indeed, that the original male members of a group who had solely accepted his entry as mate for those ineligible females, their sisters, would consent to his further participation in marital right with other female group members? Evidently not: for thus the sexual prerogatives of the strangers would be much greater than their own—for the resident males are barred by the primal law from the wives of the new comers, who yet, as resident females, form probably much the most numerous section of the feminine element in the horde. If the new comers further inherited the ordinary right of intercourse with their own daughters, who would be correspondingly numerous, then the extent of their rights would entirely outbalance that of their brothers-in-law. As original residents the latter would, however, be the law-makers, and we can have no doubt as to what form in such a case law would take.[293] Thus is struck a blow again, however indirect, to incest as a custom, a blow whose power would be the more effective, insomuch as here it is the living father himself, in the outside suitor, who would be in cause. But even admitting that it is possible to conceive a complacency in regard to such participation in sexual rights on the part of the brother, there would still be another much more formidable obstacle to incestuous license as regards his daughter confronting the male intruder in the person of the precedent sister, now his wife.

A psychological factor of enormous power was now for the first time in the history of the world to make itself felt. It would be the play of the natural feeling of sexual jealousy on the part of his resident female mate. The jealousy of a woman, in fact, is at length able to make its strength appear, to some purpose. As a wife who had not been captured, who, in fact, as an actual member of the group itself, was, so to speak, the capturer, her position in regard to her dependent husband would be profoundly modified in comparison with that of the ordinary captive female. Whereas such a captive, seized by the usual process of hostile capture, had been a mere chattel utterly without power; she, as a free agent in her own home, with her will backed by that of her brothers, could impose law on her subject spouse, and such law dictated by jealousy would undoubtedly ordain a bar to intercourse between him and her more youthful, and hence more attractive, daughter.

By these then, and other incidents, each of vast value, we may perceive how the primitive mind became gradually prepared for a change so imperatively necessary for all future progress, and how a habit even so deeply ingrafted as incest may primarily have been forced to slacken its hold. It is even possible to imagine how from

---

[293] All the younger generation of females would be reserved for themselves, and thus not only their own daughters, but the daughters of their brothers-in-law, who, as of the same generation, were all classed together as sisters.

such a point of departure, the custom might at once have entirely ceased among all, or at least a portion of, mankind. If we could conceive at this stage a secession from their original group of its resident component females, accompanied by their outside mates, with a continuance of the acceptance of the subordinate strange suitors in future generations of the new colony, then we could admit the probability of rapid evolution in approach to a well-known actual group formation. The persistent importation of the always dependent outsider would accentuate the movement already begun against incest—with two such associations in unison, cousinship would be recognised, and peaceful connubium in 'cross-cousin' marriage between groups would become a habit, and female descent the rule.[294] But at such a stage in social evolution, it is impossible to accept the dominance of the unsupported female or 'feme sole.' Gynæcocracy, if it has indeed ever existed, is evidently as yet incredible. Not thus was dealt the final fatal blow at this last great trait of archaism. We must rather seek it in the familiar economy of the type of group we have left, which is characterised, as with other animals, by the predominance of the male.

In our study of the various incidents in primitive social economy which would have had effect in a sense inimical to the custom of incest, we have only considered the matter from the point of view of the entry of the outside suitor after the death of the paternal tyrant. The incestuous rights of the living group-fathers are thereby in no way directly affected. In the absence of any direct personal interest in the matter on the part of the group-sons, the only other male components and law-makers might indeed continue to remain unopposed indefinitely. Thus a resolution of the problem of decay of incest would seem as far off as ever. Happily this is not in reality the case—the real significance of the entry of the outsider, even on such terms as we have examined, lay in the co-ordination of movement of these resultant primary checks, and the inevitable synchronous evolution of the most characteristic feature of the next and higher type of group, as in itself a mere component of a tribe. The outsider's admission, in fact, really contained the germ of progress in group formation which was to entail the total required decay [of incest].

Up to the present, although the entire male element in a group was divided into two classes, by generations whose interests had little in common, between them no antagonism had arisen which could not be appeased by the evolution of such a law in remedy as we have noted. The case would now be altered; an

---

[294] These groups would be phratries, or the germs of phratries.—A. L.

irreconcilable breach was about to divide them. It will be seen that the advent of the outsider had been a real portent. Where, for instance, and under the circumstances we have portrayed, he had become a more accustomed figure as an immigrant, he would form a valuable connecting link between groups. Each would certainly possess some females seized from the other by more or less forcible capture, but each now possessed a certain proportion of these males, brothers of those females, whose intrusion had been peaceably accepted. With less strained relations and greater intercourse, capture would become a little more rare, and a friendly interchange of women more common. There would be then discovered by the brother a hitherto undreamt-of virtue in the young female, his sister; in fact, her value as a negotiable article would appear.

As brothers and sisters, and thus barred in union by the primal law, their relative interest in each other had been of the feeblest in the past. The ultimate destiny of the sister might be a matter of the most perfect indifference to the brother. With the new order of things she had suddenly become more precious. As an object of barter for the sister of another man, she would show herself to be invaluable. In view of the difficulty and danger attending hostile capture, the temptation to such easy procuration of mates as sister-barter offered would be irresistible. Coming at first into practice when only the death of the father had left his widowed daughter free, its advantages to the sons would impose a gradual encroachment on the rights of possession by the living parent. In prejudice to incest were now opposed the two most powerful passions in human nature, sexual desire, and a love of material gain, and the successful barter of a sister for another man's sister satisfied both. For attempts at capture might be unsuccessful, and purchase might be more or less unsatisfactory. And these passions would be aroused in bosoms able to make their power felt. The sons, as also resident males, would be among the law-makers. However powerful the father in past authority and tradition, in the end the force of numbers would tell. However numerous the group of fathers, they would always be outnumbered by the group of brother sons, and victory would thus ultimately incline to these.[295] However long and doubtful the struggle, as the latter possessed the longer lease of life, the quantum of the exchange value of a sister would always finally be made to show itself, and the determination

---

[295] The breach between father and sons could only be healed by the submission of the fathers. Then prerogative in incest would gradually decay, for strange to say no vestige of law in avoidance can be traced.

to profit thereby would be more strongly impressed on each generation.

Natural selection would again certainly come into play in favour of such groups, thus curtailing the monstrous prerogatives of the old-world fathers, by dint of numbers alone. The superiority which would ensue with each generation, would speedily ensure the triumph of that assembly which could definitely accept the presence of the outside suitor. He would come as a multiple unit of strength, a willing ally who would otherwise have been an active enemy—the generator of the productive power to females who would either have remained as sterile residents, or seceded to hostile hordes as breeders of new foes.

Thus, then, we may at length perceive how a custom even so deeply ingrained in nascent man as paternal incest, may finally have become extinct as a custom. In the action of such circumstances we can accept the idea of its ultimate decay and death. By the numerical preponderance of the individuals within a group interested in its disappearance, was alone such a result feasible. This necessary condition we here find fulfilled.

In opposition to the father we now see arrayed not only the wife-mother jealous of her mate, not only the daughter inclined instinctively to youth and the unknown, but, most important of all, the son, now egged on by most powerful personal feelings and interests. And for these latter ones, as we have seen, time itself would fight; to youth each hour and day is a gain in strength, to old age each moment means a loss of power. With the decay of the custom we see that the way lies clear to progress in group formation. Sooner or later the presence of the offspring of the outside suitors in the formerly purely consanguine circle will be recognised, their recognition as cousins to the younger resident members will be made, and the old type of horde by a process of cleavage divides itself into two intermarriageable clans, (phratries?), and the savage tribe is created.[296]

---

[296] It will be observed that Mr. Atkinson, when he writes of 'the cleavage of the old type of horde into two intermarriageable clans, creating the tribe,' differs from the opinions already expressed by his editor. By 'clans' Mr. Atkinson here means 'phratries,' and we have shown that phratries, even now, often bear totemic names, and probably were, in origin, local totem groups; each containing members (by female descent) of several other totem groups. Mr. Atkinson, as far as his MS. goes, appears to have given no attention to the origin and evolution of totem names, totem groups, and totem kins. Thus he writes, 'the presence of the offspring of the outside suitor in the formerly purely consanguine circle will be recognised.' But if

Thus did the custom of paternal incest disappear, and so completely as not to leave a trace of its passage in recognised usage among actual peoples. But as an unauthorised habit it long existed, nay, it still lurks, and as such it is probably much more common among the lowest classes of even most civilised peoples than is generally imagined. The continual domiciliary propinquity of such close relatives makes the crime easy[297] and detection difficult. Amidst the savage races, although rare, it is by no means unknown.

---

the heterogeneity in the circle was only recognised as marked by female descent, and by the totem name of the female mate from without, male parentage of 'the children of the outside suitor' would not come into the purview of customary laws, would not cause the 'cleavage into two intermarriageable clans,' or 'phratries.' There was no such 'cleavage,' as we have argued, and the permission of cross-cousin marriage is due (I suspect), not to such early legal recognition of male descent, but simply to the natural working of the totemic exogamy, plus female descent.

Mr. Atkinson's theory of 'cleavage,' it will be remarked, does not involve the idea that the members of an 'undivided commune,' being pricked in conscience, bisected it for reformatory purposes. He merely suggests that his clients found, in their group, persons marriageable according to their existing rules of the game, and married them. But these persons are, according to him, recognised as the offspring of 'outside suitors' male, and are also recognised as cousins, on the female side, though even now no name for cousins exists in Australian society. This involves counting both on the male and female sides, which, in practice, may have occurred. But the theory of Mr. Atkinson avoids all the problems of the different totemic names given both to the born members of his original group, and to other members thereof, consisting of the offspring of the outside suitors. If totemic group names already existed, these suitors must have been of many totem group names. Whence, then, came the two different and distinct totemic group names of the two sets of cross-cousins—now phratries on Mr. Atkinson's theory?

Give his original group a name, say Emu. With Totemism it will contain captive wives of various groups, say Bat, Cat, Rat. It will also contain outside suitors, probably of the same names. These men are allowed to marry women of the group, and, by Mr. Atkinson's theory, the offspring of these unions, or 'cross-cousins,' are allowed to marry the children of their aunts within the group. There are thus, within the group, two intermarrying 'sides of the house,' veve, as in Melanesia. But why or how do these sides of the house, practically phratries, now receive totemic names, say Yungaru and Wutaru, or Wolf and Raven? Perhaps Mr. Atkinson would have replied, 'by a mere extension of the habit of adopting totemic names,' which, of course, involves the pre-existence of that habit.— A. L.

[297] But not tempting, according to Dr. Westermarck!—A. L.

It is not a crime by the laws of totem kinship with female descent, the daughter in such a case being always of the same totem as her mother, and thus theoretically eligible. The only bar is the classificatory system which, based on sequence in birth, forbids all connection between those of different generations. Thus this form of incest, when it does occur, in no way creates the utter horror which we find universal at any union between brother and sister. An old native chief whom I questioned on the matter certainly spat with disgust at the idea, but again, to my own knowledge, a case occurred where a girl bore a child to her own father, and when the fact was mentioned among the people, it only caused coarse laughter. It is true that in this case the culprit was a great chief—it is possible that there would have been more adverse comment if he had been a commoner. It is certain that the betrayal of the vested interests of the future husband (for in New Caledonia all children are betrothed at a very early age) would have been more resented in the latter case. But license in sexual intercourse within forbidden relationships seems everywhere the privilege of irresponsible rank, if we may judge by the Kalmuck proverb, 'Great folk and the beasts marry where they please.'

However, its occurrence in such cases may be traced to sources which show that here the exception proves the rule. Indeed, the fact of its occurring almost solely among the higher classes [as among the Incas], points clearly to a probable connection with an idea of pride of race, or a question of inheritance. Now we may note that with descent in the female line the right of direct succession to the paternal name, or place of power, or property, is not in the gift of a father. The only legal conveyers of the blood right within him are females in whose veins is to be found that same blood, i.e. his mother and sisters. However regal a personage his child by a foreign woman, it is cut off from that heritage, nor in connection with this offspring can pride of race find a place. Thus, then, we may understand how union, although illegal, with a sister was so frequent in, and even enjoined on, the royal race amidst certain peoples. The purity of the royal blood thus alone remained intact, and from a king was born a king. For it is a remarkable fact which must be more than a coincidence that amongst these very peoples, such as the ancient Egyptians, Persians, and Peruvians, whose rulers were addicted to the habit, female descent was the custom [?]. At least I am not personally acquainted with any exception to the rule. In consonance with this descent through females only and where any approach has been made to gynæcocracy, we shall expect to find that there would be only one legal wife. Such was indeed the

206

case also in Ancient Egypt, there is no instance of two consorts given in any of the inscriptions. This fact, taken in connection with that which conduced to incestuous union under this form of descent, invites us to make a digression in a curious reflection, not however entirely foreign to our general theme. For the same effect as regards inheritance on the offspring which would be produced by union with a sister, would also occur in marriage with a daughter whose parents had been themselves brother and sister. Thus we may guess the lineage of the unknown mother of the great royal wife Nefer-ari, daughter and consort of the Pharaoh of the oppressive Rameses the Great. This daughter had, in fact, been probably chosen among others for wife precisely because her mother had herself been both his sister and his wife.[298]

We may now renew our affirmation that paternal incest as a custom, is no longer generally recognised anywhere. The primitive unquestioned marital right in incest is quite unknown. It has disappeared, and so completely have even traces of its past general occurrence faded, that doubts of the reality of the fact may be pardonably entertained. The question is of importance in connection with our thesis, for as may be seen the whole theory of the primal law is based on the idea of its primitive universal prevalence. We hope, however, to have shown the inherent possibility of the fact as being a habit common to all the mammalia—and it has seemed against reason to suppose that man's ancestor, whilst in the animal stage, would be an exception to so general a rule. Our further argument has adduced circumstances in favour of a final decay so complete that oblivion could not but follow.

Perhaps not the least remarkable fact to the anthropologist in connection with its life and death, is that only as between a father and a daughter, of all blood relationships, do we find no trace among actual peoples of any law in Avoidance. The fact is significant, as we may thus surmise that the process of decay was very long delayed, in fact to a time when such inchoate form of law as Avoidance had become an archaism, or until general progress had rendered any law unnecessary.

---

[298] It will be interesting to see if research will bring to light the fact that even with so irresponsible and imperious a dynasty as the Ramesids some form of lustration was not considered necessary in the event of such unions. This is the case with the people of Madagascar under similar circumstances.

CHAPTER VII

TRACES OF PERIOD OF TRANSITION— AVOIDANCES

Survivals in custom testify to a long period of transition from group to tribe.—Stealthy meetings of husband and wife.— Examples.—Evidence to a past of jealousy of incestuous group sire.—Evidence from Teknonymy.—Husband named as father of his child.—Formal capture as a symbol of legal marriage.—Avoidance between father-in-law and son-in-law.—Arose in stage of transition.—Causes of mother-in-law and son-in-law avoidance.— Influence of jealousy.—Examples.—Mr. Tylor's statistics.— Resentment of capture not primal cause of this avoidance. —Note on avoidance.

With a custom so deeply ingrained as incest would be in the nature of man's ancestor, still doubtless vastly animal, we may indeed surmise that the process of its decay was long and tedious. The temptation, as we have said, to such easy procuration of a mate in comparison with the danger and comparatively scanty results of capture, was very great, whilst the continual propinquity of father and daughter would tend to constant recrudescence, especially in default of any trace of law against it. There must, then, evidently have been a transition era of vast durance, between the type of the isolated consanguine group whose only resource in matrimony was exogamic hostile capture, as the outcome of the incestuous lust of its solitary male head, and the all-embracing tribe composed of an aggregation of several groups, and possessing thus ipso facto all the necessary elements of an endogamic connubium quite incompatible with such incest. In such a tribe, a group of women in many cases formed the pivotal centre, and capture was often found only as a form in survival. Is it possible to retrace the main features of an epoch of such evident importance in social evolution? In view of the fact that, in the past course of our argument, such law as would seemingly have been necessarily evolved in regulation of each step in primitive progress has been found identical in form with some actual savage custom, may not a deeper investigation of savage custom disclose further co-ordination, and prove equally fertile in interpretation of the past? Whilst, again, many obscure observances in actual lower life, in consonance with such archaic genesis, may take a rational form, though the origin seems apparently lost for ever.

Such research will, I think, clearly show that many social features in modern savage habit afford internal evidence that, as a fact, they could only have arisen in such a transition era. They also bear marks of a very lengthened evolutionary process, and thus confirm the natural idea of a halt of portentous length at the threshold of the present haven of comparative social rest. We shall doubtless find that the door left ajar by the entrance of the outside suitor was not to yield further with ease to the pressure of new needs, half-hearted as men would be, from the conservative force of old ideas, of incest and entire masculine dominance.

There is, for instance, one curious trait in actual savage custom which evidently dates from a very early stage of this epoch. It is that of the strange forms of 'stealthy' intercourse, being the indispensable preliminary symbols of the legality of an after marriage between the resident female of a group and an outside male. These forms are well known to anthropologists as occurring among many lower peoples. Here we find that the visits of the male suitor are supposed to be distinctly clandestine, taking place only by night, although in reality the fact is perfectly in the cognisance of the whole group. Now such fugitive and secret meetings are exactly what would have taken place when a group had arrived at a stage in which, although filial incest was decaying as a custom, there were still recognised certain marital rights over his daughter by the living father; when, in fact, tolerance of the presence of the outsider was yet in a tentative stage—and he was still regarded with suspicion, if not disfavour.[299]

In consonance with the view we have advanced of the circumstances attending the entry of an immigrant suitor, it has seemed to ensue that his position would have been quite dependent, and himself considered as a foreign element. That such was actually the case seems again proved by another trait in modern custom, whose genesis, however, was of very much later date, and when speech had made some progress. In our own day clandestine intercourse, as above described, may continue to pregnancy. On the birth of the child alone does the father become recognised as part of the group. But even so his nomenclatory power as regards his offspring is absolutely nil. Far from giving a name to the child, his own is taken from his offspring. Till now, in fact, he has been

---

[299] Well-known instances of this marital shyness are the Spartan and Red Indian usage of only entering the wife's bower, or wigwam, under cover of darkness. There are also Fijian and New Caledonian cases (Crawley, pp. 39-40). Mr. Crawley would regard these as cases of 'sexual tabu,' but various other cases may be readily conjectured.—A. L.

nameless; in future he will be known as the father of so-and-so, of Telemachus, in the case of Odysseus. To this point we will, however, have to return when we arrive at the question of the evolution of personal descent from that of descent recognised by locality, which we consider to have been the most primitive form. [Mr. Atkinson probably means descent from a local group, say Crow, not descent denoted by a place name, as 'de Rutherford.']

There is another trait in actual custom which also could only have acquired its most remarkable features in this era of change, and that is hostile capture itself, in its legacy of those 'forms' of capture which we find connected with more peaceful connubium all over the world. Such 'forms' have rightly been considered as mere survivals, and thus in agreement with our own theory capture is generally accepted as the earliest form of outside marriage.[300] But in some minds the brutality necessarily attending real capture, and its occurrence solely among very low races with whom any idea of sexual restraint is expected to be quite unfamiliar, has simply connected the process with the general lawlessness which, amongst such peoples, is supposed to characterise the relations between the sexes. Its occurrence in form of survival among higher races has been considered a meaningless ceremony, and its evident symbolism in legality dismissed as incredible. Students are, however, aware how much in error is the idea of utter lawlessness in connection with the marriage relationships of any savage race. On the contrary, as is well known, the list of prohibited kindred is not only much wider than our own, but no stage in the marital arrangements is without irksome and minute legislative restraints, strictly limiting and defining the rights of each individual, male and female.

To other minds the fact that a 'hostile capture,' presenting as its most characteristic feature an utter violence, should ever have been constituted into a symbol of legality in marriage, has given rise to much perplexity. Mr. McLennan in fact remarks—'It is impossible to believe that the mere lawlessness of savages should be consecrated into a legal symbol'—an assertion which we may accept, however little we are prepared to accept his general views on early society. It is evident that the whole difficulty has arisen from the apparent complete incompatibility of a seeming method in violence with a virtual act in law. The hypothesis we have presented of the 'primal law,' and its exogamous sequel, would seem however to throw a new light on the matter. All unions within the group being by the action of primal law, as we have shown, considered

---

[300] See Note at the end of chap. V.

incestuous and illicit, marriage could only take place with an outside mate. The presence of a captured female within the camp would thus, as we see, actually constitute in itself a proof, and the only one possible at the epoch, of the legal consummation of marriage as ordained by the primal law. It is thus easy to see how a form of capture should be retained as a symbol of legality in later connubium. Its continued vitality results from the intense conservatism of lower peoples, and from the fact of the halo of prowess that surrounds it.

Its evolution as a symbol only arose, however, when, during the transition era, by the conjunction of groups into a tribe, friendly unions were possible. It would not have occurred with the earliest forms of horde, for these were isolated and hostile, and real capture itself was the sole form of marriage; nor would it have occurred with that later type, in which, with matriarchal descent, the relative positions of males and females were reversed, as far at least as suit in union is concerned.[301] It took its rise with that other great type of group characterised by patriarchal descent, which in all the after history of the world (for, as we shall see, their evolution was coincident and had for cause the same factor) was to dispute supremacy with that which accepted uterine descent. Here, as in the original type, the male continued to preserve his predominance and continued its traditions of capture.[302]

There remain other actual traits whose connection with this era is equally evident. For instance, avoidance between father-in-law and son-in-law could not have had its genesis in the very earliest type of assembly. Whilst parental incest ruled as the custom, each group must have been isolated from and hostile to every other. These two could never have been in habitual presence one of the other. But, again, the habit could not have arisen in the later form, as represented by a tribal horde with uterine descent, as

---

[301] With the consequent accession of power to the resident female thus accruing, capture would have become more rare. In any case it would certainly become connected in the minds of the more advanced and powerful tribes with the rape of women, other than their own, and probably inferior in type, mentally and physically; the comparison of this degraded captive in their midst with their own free females would not be at all likely to have led in connection with her to any spontaneous idea of symbolic consecration in marriage, or aught else.

[302] When two groups, despite the isolating tendency of the habit of capture, did at length form a union sufficiently close to permit of marriage by consent between the respective group members, then, with capture as regards outsiders still rife amongst them, we can understand how the symbol would come to be attached to the peaceful connubium.

primitively composed of only two intermarrying groups, each of which formed a clan distinguished by a different totem emblem.[303] The relative clan-relationship of each member of the horde would, by the aid of this distinctive totem, be distinctly defined, and, with female descent, the father-in-law and son-in-law would find themselves members of the same clan [phratry]. As thus being both males and of the same 'phratry,' there could not possibly be avoidance or enmity, real or simulated, between them.[304] By all the sacred ties of blood [phratry] they were conjoined in offence and in defence. Further, where descent is uterine we find that the disposal of a daughter is in the hands of the mother or maternal uncle alone—the father has no voice whatever in the question, nor any part in her value as an object of barter or sale. Thus he is perfectly disinterested in the matter of his children. So far from being in disunion with his son-in-law, his sympathies, in case of a tribal quarrel, would be certainly with him. But the younger man, in internal quarrels, might be found fighting to the death with his own real father, not (as I have seen it stated in mention of just such an incident)[305] because he has become part of his wife's clan,[306] which could never be, but because, with descent through the female, his father would be a member of the different group and of other blood to himself, and to his father-in-law also.

The genesis of this particular avoidance (father-in-law and son-in-law) took place during that stage of the transition era, when, incest still lingering, the immigrant suitor was so far acknowledged that his entry into a group was not always delayed till the death of his proposed father-in-law. As they were thus possible rivals there was a chance of friction, only to be averted by the law in question. Avoidance would arise at the same time between mother-in-law and son-in-law, but this time as a measure of protection for the marital rights of the husband of the former.[307] It could not have arisen in the early Cyclopean era. The son-in-law as such, could evidently not have had existence when the mother's daughter was the father's wife, nor, later, when, with the general recognition of the

---

[303] 'Phratries' are here meant, where the word 'clan' is used, or local totem groups.—A. L. Cf. Note, p. 260.

[304] The exact relation of each to the females being defined by the classificatory system by generations.

[305] As mentioned by Tylor.

[306] Here I really do not know what 'clan' is meant to denote—'phratry,' I think.—A. L.

[307] See Mr. Crawley's 'Sexual tabu' theory of this avoidance, Mystic Rose, pp. 399-414.—A. L.

classifactory system, there arose a strict interdiction of sexual union between members of different generations. There would in such circumstances be no further risk of danger from the jealousy of a father as regards his wife, and the husband of his daughter. It had its origin in the fact that when the outside suitor had originally been granted entry, it would only have been after the death of the patriarch sire, and as a mate for his widowed females. But as these would include both mother and daughter, there would thereby be created a precedent, so to speak, which required regulation, when later, with the decay of incest, the living father remained in presence. In fact, avoidance between mothers-in-law and sons-in-law defined fathers-in-law's rights.

We may here again note another step in advance to purely human attributes in the fresh distinction between female and female, which has now again arisen as between a mother and her daughter as regards the immigrant suitor. But whereas with these, as indeed with most of the cases of avoidance we have studied, sexual jealousy has been the primary cause, we may now trace the action of quite another factor, which would certainly tend to a conservation of the habit, and in a manner intensify it. This would be association of idea with hostile capture.

As regards the father-in-law, however, the custom, as far as capture is concerned, would not occur with female descent, for the reasons we have already given of clan kinship in such a case. It might, however, be found as a factor to a certain extent with male kinship, for here it is the father-in-law who is of the same clan as his daughter, and thus interested in her negotiable value. Thus it is possible to imagine enmity between him and her possible captor, who is also of a different clan from himself. As regards avoidance of mother-in-law it has again, perhaps, been accentuated by forcible capture. The effect, however, in relation to descent would be exactly the reverse of that with the father. With early male descent in the primitive tribe as composed of only two clan groups [phratries], it is she who would be of the same stock as her daughter's husband, and the habit would not arise, the captor is, in fact, a member of the tribe from which she herself has been stolen; although later, when more than two clans were conjoined,[308] it might happen that her son-in-law belonged to another, and here there might arise feelings of animosity. With uterine descent the case is certainly altered. As a mother, and as member of a clan different from that of the male suitor, the figure of the son-in-law might be dreaded as a possible

_____

[308] Apparently 'clans' here = totem kins, Mr. Atkinson seems to think that totem kins kept on being added to the two original 'phratries.'—A. L.

captor of her daughter and other young female members. But here again a difficulty arises, for when the capture becomes an accomplished fact, the mother-in-law and son-in-law would probably not meet again, at least in primitive times: he belonging to the group having patriarchal descent, with capture as the rule; she, to the matriarchal, where the female is normally immobile, between which two forms of group no friendly intercourse could occur. The fact of avoidance in any form presumes contiguity or the habitual presence of the individuals concerned, and this in such a case could not arise. So as in Tylor's figures we find that in W to H, as the latter is completely cut off from his family, there is not one single case of avoidance between the wife and the husband's relatives. It is evident that the same rarity of contiguity must have arisen also with the father in male descent; there is here certainly cause of disagreement in the rape, but if the parties see each other no more there would be no necessity of evolution of avoidance to mark the fact. Indeed, I cannot help thinking that the importance of association with hostile capture has been much exaggerated as a factor in the evolution of Avoidance. The question of 'residence' and 'descent' has not been held sufficiently in account by those who insist on the capture as the sole cause of avoidance. Despite the eminence of the authors favouring this view, I would venture to submit that the balance of proof would much favour sexual jealousy, which we have heretofore found the sole motive power in all changes.

Those who would uphold anger roused by capture as the cause of avoidance with the wife's relatives, for instance, must be prepared to show that it would be strongest with the one who was most deeply interested in the wife, one whose voice in her destiny was of greater power than her own mother's, and that was her maternal uncle, the head of her clan. Now I have failed as yet to find a single trace of such a case as avoidance between the latter and his sister's daughter's husband.

Again, jealousy, or a desire for regulations in matters of sexual union, will explain certain details in the accounts we have received of individual cases which seem otherwise obscure or irrelevant. These have been overlooked, as they are minute, but from my point of view are full of significance when closely examined. Mr. Lubbock says,[309] quoting Franklin as to American Indians: 'It is extremely improper for a mother-in-law to speak or even look at him, i.e. her son-in-law.' Quoting Baegert: 'The son-in-law was not allowed for some time to look into the face of his mother-in-law.' Further, 'among the Mongols a woman must not speak to her father-in-law,

---

[309] Lubbock, Origin of Civilisation, p. 13 et seq.

nor sit down in his presence. Among the Ostiaks, Une fille mariée évite autant qu'il lui est possible la présence du père de son mari tant qu'elle n'a pas d'enfant, et le mari pendant ce temps n'ose pas paraître devant la mère de sa femme' (Pallas). In China the father-in-law after the wedding-day never sees the face of his daughter-in-law again, he never visits her, and if they chance to meet he hides himself. Among the Kaffirs a married woman is required to hlonipa[310] her father-in-law, and all her husband's male relations in the ascending line, i.e. to be cut off from all intercourse with them.

Again, in Australia, it is compulsory on the mothers-in-law to avoid the sight of their sons-in-law, by-making the former take a very circuitous route on all occasions, to avoid being seen, and they hide the face or figure with the rug which the female carries with her. So strict is the rule, that if married men are jealous of any one, they sometimes promise to give him a daughter in marriage. This places the married man's wife, according to custom, in the position of mother-in-law, and renders any communication between her and her future son-in-law a capital crime.[311] Also among the Sioux or Dacotas, Mr. Philander Prescott remarks on the fear of uttering certain names. The father and mother-in-law must not call their son-in-law by name, and vice versa, and there are other relationships to which the prohibition applies. He has known an infringement of this rule punished by cutting the offender's clothes off his back and throwing them away. Harmon says 'that among the Indians east of the Rocky Mountains it is indecent for the father or mother-in-law to look at or speak to the son or daughter-in-law.' Among the Yakuts, Adolf Erman noticed a more peculiar custom. As in other northern regions the custom of wearing but little clothing in the hot stifling interior of the huts is common there, and the women often go about their domestic work stripped to the waist, nor do they object to this disarray in the presence of strangers; but there are two persons before whom a Yakut woman must not appear in this guise, her father-in-law and her husband's elder brother. Again, quoting J. G. Wood, he says the native term for these customs of avoidance is, 'being ashamed of the mother-in-law.' The Basuto custom forbids a wife to look in the face of her father-in-law till the birth of her first child—and among the Banyai a man must sit with his knees bent in presence of his mother-in-law, and must not put out his feet towards her.

---

[310] Hlonipa, to avoid mention of his name, &c.
[311] Origin of Civilisation, p. 14. Lubbock quoting 'Report of Select Committee on Aborigines,' Vict. 1859, p. 73. Tylor, Early History of Mankind, p. 288.

Now an important circumstance to be remarked in nearly all cases of Avoidance is, that it principally exists between people of different sexes,—thus an a priori inference may be drawn that the primary cause lay in some relation to the sexual question. It is significant that a woman's avoidance of her husband's relations is with those in the ascending line, i.e. with his seniors. Against his juniors he can defend himself, against his seniors he needs the protection of law. In the cases we have cited, it is significant that, besides the father-in-law, hlonipaed among the Kaffirs, the woman must hlonipa all her husband's male relations in the ascending line. Among the Yakuts she must not appear unclothed before her husband's elder brother.

Among the Veddahs of Ceylon a father will not see his daughter, nor a mother her son, after they have come to years of maturity.[312]

If we examine the words italicised in the quotations above, they seem to convey more nearly an idea of impropriety in any approach to intimacy than that of 'cutting' from enmity, as Dr. Tylor has suggested. Indeed, we observe here just the same horror that a too familiar attitude between forbidden kindred, as uncle and niece, would excite amongst ourselves, arising from the same idea of repugnance.

We see that various observers use the terms 'improper' (Franklin), 'the fear of' (P. Prescott), 'indecent' (Harmon), 'cut off from all intercourse with them,' and no doubt they have[313] each expressed the impression made on themselves in observation. We note again that the only case where the native term in designation of the custom is given that it means 'being ashamed.' The limit in time for the avoidance is again significant. 'For some time,' Baegart says; 'tant qu'elle n'a pas d'enfant' (Pallas). 'Till the birth of the first child.' These limitations in time would not exist if enmity because of capture was the cause, whereas we can quite understand them if, the circumstances now proving the consummation of marriage, jealousy might then be supposed to cease. The reserve as to a too familiar attitude that this idea of indecency would imply, is shown where a Mongol daughter-in-law 'cannot sit down in the presence of the father-in-law,' and where the Banyai man 'must not put out his feet towards his mother-in-law, but sit with his knees bent in her presence.' In China it is the father-in-law who hides himself, and

---

[312] Among the Veddahs the fact that the avoidance begins after puberty, and in each case in relation to the opposite sex, is evidence that here the sexual feelings are concerned.

[313] Tylor, Early History of Mankind, p. 291.

this surely would hardly be the act of a captor, nor can we imagine a man having his clothes cut off his back simply because he had not 'cut' some one sufficiently.

However, in connection with our argument we have Adolf Erman's account of the custom among the Yakuts, and where we find the husbands elder brother joined with the father-in-law in an avoidance, there a distinct feeling of impropriety in connection with these relations in law of the wife is indicated. The diffidence cited is exactly what would occur if union was undesirable and yet not impossible, between the persons in avoidance, and hence temptation was to be avoided. It is very important to note that no idea of enmity from capture can be associated with the husbands elder brother. Again, the custom of avoidance with an elder brother, where its connection with jealousy is evident, is very widespread, and very strict in observance; as we have already noted, it occurs in Orissa and among the Kyonthas in India, whilst I have also observed it in practice in New Caledonia, where it is most undoubtedly a means to an end, to protect the younger brother's marital rights. As to the significance of the fact mentioned in the case of the natives of Australia, where, as regards their wives, they are jealous of a man— and give him a daughter to place him in avoidance with her mother, comment is unnecessary.

These facts seem to me to be conclusive; but the question of the exact origin of avoidance is so important to my general argument, that I am glad to be able to find what I fancy is added proof from another source. If this furnishes the requisite evidence that sexual jealousy was the real factor, and not hostile capture, our hypothesis of the primal law acquires valuable inferential evidence in its favour. Such added proof we hope to be able to show in Dr. Tylor's figures.[314]

	----	Theory as	practice	--
In Residence, H to W: 65				
Avoidance H to W relations	(a) 9 cases	(b) 14 cases		(A)
Avoidance W to H relations	3 "	0 "		(B)
In Residence, W to H:				
Avoidance W to H relations	5 "	8 "		(C)
Avoidance H to W relations	18 "	9 "		(D)

These figures, which are extracted from Dr. Tylor's work,

---

314 E. B. Tylor. On a method of investigating the development of institutions: applied to laws of marriage and descent. J. A. I. 1889, xviii. No. 3, 245-269.

would seem to be eloquent against hostile capture being the sole cause of Avoidance. They are derived from a comparison of Avoidance as occurring, (a) quite independent of residence, and (b) as actually resulting where coincidence of Avoidance and residence is found.

Now as regards the question of jealousy as cause of Avoidance, residence and propinquity will evidently have a powerful effect.

(A) As we have seen, any Avoidance under these circumstances would be remarkable without a prior stage in quite other conditions than those found generally with H to W residence. We note that whereas we might expect under even the above conditions to find only 9, there are 14. Here sexual jealousy has been an important cause.

The Avoidance of the Mother-in-law (for, of course, there was none here with father-in-law, who was a nonentity in such a family circle, and of the same clan as the son-in-law) arose as a matter of protection for the marital rights of the daughter as against her mother, both inhabiting the same large house common to matriarchal descent.

(B) Here, again, we expect to find 3, and see there are actually none, from which it would seem to result that W capture had nothing whatever to do with the origin of A, H to W, for, admitting the almost entire separation of the W from H family, which would make the case rarer, a tradition of capture would exist which would have effect when they were later grouped together. Whereas the non-Avoidance is explained by lack of jealousy, from absence of male relations of H.

(C) Here it is again quite impossible to accept any idea of W capture as the motive cause. Avoidance arose between W and father-in-law to protect rights of son-in-law and mother-in-law. It was evolved, as we have seen, as a measure of protection for that generation of males who were the actual captors, each generation by the classificatory system having individual rights. That the necessity for such legislation was urgent we see in the proportion of the figures 5 to 8.

Here, again, the fallacy of capture as primal cause of Avoidance is clearly evident. If this was the case, we might expect it to be almost universal, whereas in reality, instead of the 18 cases which the average should give us, we find only 9. It really had its origin in the reason we have already given, of sexual jealousy as a primary cause, and was later augmented as serving to impress on many the classificatory distinction between M and D, who otherwise, as far as totems went, were eligible to the same person.

Where both father-in-law and mother-in-law are in avoidance, we may surmise a change in descent from the F to the M in the tribe, the converse change of M to F of course never occurring. The question of change of descent will explain problems in the nomenclature of Morgan's tables as regards nephews and sons, which have been overlooked.[315]

NOTE TO CHAPTER VII

Mr. Crawley reckons three interpretations of the origin of the avoidance of mother-in-law and son-in-law. 1. Fison (Kamilaroi and Kurnai, p. 103), 'It is that the rule is due to a fear of intercourse which is unlawful, though theoretically allowed on some classificatory systems.' Mr. Crawley remarks, 'this explanation is the one most likely to occur to explorers who have personal knowledge of savages,' which was Mr. Atkinson's case. Mr. Crawley objects the antecedent improbability of any man, 'not to mention a savage, ever falling in love with a woman old enough to be his mother or mother-in-law, and the improbability of so many peoples being afraid of this.' Now 'in love' is one thing, and an access of lust is another. Moreover, the mother-in-law, in prospective, not infrequently is her daughter's rival, even in modern life. She has to be guarded against, even if the son-in-law is less dangerous. And he is very apt to be 'a general lover.' 'Theoretically the mother-in-law is marriageable in many systems,' says Mr. Crawley, 'and so there would be no incest....' But Mr. Atkinson is not contemplating the danger of incest as the cause of mother-in-law avoidance; his theory postulates jealousy—that of the mother-in-law's husband, and, for what it is worth, that of the mother-in-law's daughter. Mr. Crawley's objection, I think, does not invalidate Mr. Atkinson's theory; especially as he does not reflect that the possible mother-in-law may have a caprice for her son-in-law, while the would-be son-in-law, less frequently, may follow the course of Colonel Henry Esmond.

2. Sir John Lubbock's (Lord Avebury's) theory, of enmity caused by capture, Mr. Atkinson has dealt with; it is rejected by Mr. Crawley.

3. Mr. Tylor's theory (Journal Anthrop. Institute, xviii. 247), is that of 'cutting' 'an outsider,' not one of the family, not recognised

---

[315] The matter here is highly technical, and must be compared, if it is to be understood, with Mr. Tylor's essay, cited in the previous note. W stands for Wife, H for Husband, D is Daughter, F is Female, M is Mother, and is also Male! A is Avoidance.—A. L.

till his first child is born. For various reasons, Mr. Crawley rejects this explanation, rightly, I venture to think. Mr. Crawley holds that the mother-in-law avoidance 'seems to be causally connected with a man's avoidance of his own wife,' which he regards as only one aspect of the tabu between the two sexes, superstitiously regarded as dangerous to each other. But, like Mr. Atkinson, I much doubt whether the 'avoidance,' as far as it goes, of husband and wife is, in the main, the result of this superstition, though it plays its part on special occasions, as before the women sow the crops, and before the men go forth to war. Mr. Crawley's suggestion that, as husband and wife are perpetually breaking the alleged sexual tabu, the mother-in-law becomes 'a substitute to receive the onus of tabu,' 'a good instance of savage make-believe' does not carry conviction. Mr. Atkinson's theory seems 'as good as a better' (Mystic Rose, pp. 400-414).—A. L.

CHAPTER VIII

THE CLASSIFICATORY SYSTEM

The classificatory system.—The author's theory is the opposite of Mr. Morgan's, of original brother and sister marriage.—That theory is based on Malayan terms of relationship.—Nephew, niece, and cousin, all named 'sons and daughters.'—This fact of nomenclature used as an argument for promiscuity.—The author's theory.—The names for relationship given as regards the group, not the individual.—The names and rules evolved in the respective interests of three generations.—They apply to food as well as to marriage.—Each generation is a strictly defined class.—Terms for relationship indicate, not kinship, but relative seniority and rights in relation to the group.—The distinction of age in generations breaks down in practice.—Methods of bilking the letter of the law.—Communal marriage.—Outside suitors and cousinage. —The fact of cousinage unperceived and unnamed.—Cousins are still called brothers and sisters; thus, when a man styles his sister's son his son, the fact does not prove, as in Mr. Morgan's theory, that his sister is his wife.—Terms of address between brothers and sisters.—And between members of the same and of different phratries.—These corroborate the author's theory.—Distinction as to sexual rights yields the classificatory system.—Progress outran recognition and verbal expression.—Errors of Mr. Morgan and Mr. McLennan.—Conclusion.—Note.—' 'Group marriage.'

In the gradual evolution of the group into the tribe during the long period of transition, the modifications in the internal organisation, which took place as the necessary result in the march in progress, should have left traces which we may also be able to follow in living custom. The immigration of the outside suitor, in its synchronism with the decay of paternal incest, must have entailed continual complications demanding regulation, and the resolution of each problem would lead to an almost mechanical step in advance. When by force of circumstances of environment or others such a step became retrograde, then we may expect an aberrant form whose very anomalism should lead to a facile recognition, and prove equally fertile in interpretation. Indeed, a curious vestige of the effect in action of the habit of incest, when brought into inevitable contact with progressive social evolution, is to be discerned in the nomenclature of that earliest phase of the classificatory system which Mr. L. H. Morgan has called the Malayan. From the general prevalence among lower races of a division into classes by generations of the members of group, and the deduction we see drawn in Ancient Society from the Hawaiian terms of relationship therein detailed, as to a previous state of general promiscuity, it will be desirable thoroughly to examine the whole question of the so-called classificatory system. It is doubly imperative in view of our own hypothesis, which, as regards the primary origin of society, may be said to be exactly the reverse of that of Mr. Morgan, in as far as the sexual inter-relations of brother and sister are concerned.

We have tried to portray the imperative evolution of a primal law as the sole possible condition of the first steps in social progress, a law which had so specially in view the bar to sexual intercourse between a brother and sister that it might, if a name for it were needed, be called the anadelphogamous law. [Mr. Atkinson wrote 'asororogamic,' which is really too impossible a word for even science to employ.] Mr. Morgan, on the contrary, says.[316] 'The primitive or consanguine family was founded upon the inter-marriage of brothers and sisters own and collateral in a group.' He adds,[317] 'The Malayan system defines the relationship that would exist in a consanguine family, and it demands the existence of such

---

[316] Ancient Society, p. 384, Lewis H. Morgan.
[317] Ibid. p. 402.

221

a family to account for its own existence.' And again,[318] 'It is impossible to explain the system as a natural growth, upon any other hypothesis than the one named, since this form of marriage alone can furnish a key to its interpretation.' He bases his argument on the fact that[319] 'under the Malayan system all consanguines, near and remote, fall within some one of the following relationships, viz. parent, child, grandparent, grandchild, brother and sister—no other blood relationships are recognised,' and says, speaking of promiscuity, that[320] 'a man calls his brother's son, his son, because his brother's wife is his wife as well as his brother's, and his sister's son is also his son because his sister is his wife.'

Now that a brother's son should be called a son is quite simple, as being a natural effect of the group marriage of brothers, the prevalence of which as a habit, and its effects, MM. Lorimer and Fison so well show among the Australians.[321] But that a sister's son should also be termed, by her brother, a 'son' is certainly a very different thing indeed, despite Mr. McLennan's and other arguments to the contrary. In this verbal detail lies the whole crux of the matter as regards Mr. Morgan. That it should have given rise to such diversity of opinion and suggested his theory of brother and sister marriage need hardly be matter of surprise. For it is at once, evident that a group holding such nomenclature ignored cousinship, even if it existed. To all later seeming my sister's son must be nephew to ego quite necessarily. That at any stage he should be unrecognised as such seems the more astonishing, as even in the very early times when totems first arose, and arose probably and precisely to distinguish cousins as such,[322] each cousin is of a different totem to the other, and thus not only eligible in marriage

---

[318] Ibid. p. 409.

[319] Ibid. p. 385.

[320] Ancient Society, p. 391, Lewis H. Morgan.

[321] Kamilaroi and Kurnai, Lorimer and Fison. Cf. note at end of chapter. I have already stated my objections to the theory of 'group marriage.'—A. L.

[322] 'Totems arose to distinguish cousins as such.' This implies that the totem name was assigned to each group for a definite social purpose, the regulation of degrees of kin. But, on any feasible theory of the 'totem' it 'came otherwise,' and was only used as a mark of kinship after it had come, just as a place name might have been used, had it been equally convenient. On the system of descent of the totem on the female side, A (man), an Emu, marries B (woman), a Kangaroo. Their sons and daughters are Kangaroos. C, one of the sisters, marries D, a Witchetty Grub, her children are Kangaroos. E, C's brother, marries F, a Frog, his children are Frogs, and may, as far as the totem rule goes, marry their cousins, C's children, who are Kangaroos.—A. L.

with another cousin, but in many lower races the born spouse each of the other. The whole question thus resolves itself into the exact value of the term we find used in the Hawaiian designation of the sister's son by her brother. Now it is important to note that two causes might have for effect the form of nomenclature in which a brother and sister each call the child a son, and thus ignore a possible cousinship. One cause is that some factor in self-interest or otherwise allowed such relationship to remain unrecognised, although existent, and another is that, as cousinship did not exist at all, there could be no recognition, or, as Mr. Morgan puts it, 'his sister's son is also his son because his sister is his wife.' To determine which is correct certainly seems difficult, and the whole thing has evidently been considered a most stubborn fact for the opponents of promiscuity.

That Mr. Morgan should have seized it in support of his theory, and that the theory should be so largely accepted, is not astonishing. Happily the great value of his ensuing argument as regards tribal development is in no way impaired if it can be shown, as we hope to do, that there is no necessity for an hypothesis of promiscuity to explain the terms in the Malayan table, which apparent need seems primarily to have led Mr. Morgan to evolve the idea of his primitive group. In fact, it becomes evident that, if we can furnish a clue as to how a sister's son came also to be a brother's son, without having recourse to the theory of an incestuous union of brothers and sisters, we at least discount the need of Mr. Morgan's 'consanguine family,' in which such incest is supposed to be a most characteristic and essential feature. We hope to prove that the terms which misled him are more apparent than real as proofs of any real affinity in blood, and that the original conception in causal connection was something quite apart.

Sir John Lubbock (Lord Avebury) has observed that the lower the milieu of a social status the less we see of the individual and the more of the group. In the case before us the individual as such does not exist at all, and there is only question of the group in its relation to its component classes. To confound one with the other led to Mr. Morgan's error.

There was much, in fact, in Mr. McLennan's shrewd remark in criticism of Mr. Morgan's theory that he did not seek the origin of the system of nomenclature in the origin of the classification of the connected persons, and that he courted failure in attempting to solve the problem by explaining the relationships comprised in the system in detail.'[323] But it seems to me that Mr. McLennan fell into

---

[323] Studies in Ancient History, McLennan, p. 269 et seq.

the same error when he contented himself with the misleading analogies which a comparison with the Nair family system presented. These, however striking, are, as we shall find, simply the result of the fact that class or communal marriage was the common trait of the polyandrous and the Cyclopean family, nor can I see that Mr. McLennan followed his own excellent advice as regards the possible identity in origin of nomenclature and classification; if he had so done, his acute mind could not have failed in a resolution of the whole problem, whereas his final resume of the argument is in terms which I profess to be quite unable to grasp.

Before entering into the matter ourselves, we must keep in mind our affirmation as to the axiom which must, in my opinion, guide us in all research into the hidden causes of early social evolution. All innovations, as we have said, in the regulation of society, all novel legislative procedure so to speak, will be found to have relation to the sexual feelings in jealousy. This already is the genesis of the primal law, and, in each case of avoidance, we have found jealousy the leading factor. It is the same in the case before us. Bearing this in mind, let us then follow Mr. McLennan's advice as to seeking the origin of the classification of connected persons. Now what would be the family economy of the primitive group, and who are its component individuals, whose interests, in sexual matters, are likely to clash, and whose mutual relationship in this respect demanded distinction in furtherance of regulation of their respective rights?

The original primitive type of family, which we have called 'the Cyclopean,' has disappeared, giving place to a higher form, which, by the inclusion of male offspring, has permitted the existence of several generations in presence. The component individuals, speaking of one sex only, would be old males, males, and young males representing three generations. It is the interests of these generations, which, in sexual matters and in choice of food, &c. would be likely to clash, for we may be sure that the seniors, as with actual savages, would desire the lion's share. Distinction then being necessary, it would naturally, as with individuals, be based on relativity of age, seniority within certain limits confering priority. Thus gradually each generation, as indeed with actual lower races, would, qua generation, come to be a distinctly defined class with certain separate rights and obligations. In this simple necessity of a classification of the connected persons, we see the origin of the classificatory system itself, as an institution. Divers interests, as between seniors and juniors, demanded strict demarcation, and the limits of a generation furnished the required lines to mark them.

The very natural distinction by relativity of age was simply, as with individuals, utilised as the requisite machinery in regulation of mutual rights of the individual himself. His rights are a matter of concern simply within his generation, in which the relation is purely paternal and communal, with the sole reservation of rights conferred by seniority.

Even when later denominative expression was given to the idea of a generation, terms almost identical of male, old male, and young male are used, as there is no desire to convey any idea of personal kinship, and there is merely in view reference to relativity of age of a class in relation to the group. Later, as Mr. McLennan says (p. 277): 'Whatever class names primitively signified, Kiki would come to mean child, Kina parent, Moopuna grandchild, Kapuna grandparent, but originally no such idea of kinship was in view.' The classificatory system evolved itself simply as the result of a desire to define certain rights, and the division by generations was the most natural and feasible for the purpose. But the very simplicity and paucity of the original terms show that it was applied to any simple group form. In fact, we are here dealing with that primitive form which bound people together, by the mere tie of residence and locality, and was purely exogamous in habit. Now when we consider that this fixed relativity of age by generation was originally evolved in view of the relations within such a family, we can imagine that complications might arise from such arbitrary definitions, when, later, this family expanded into the numerically large tribe composed of two intermarrying totem clan groups [phratries].

Primitively, doubtless, as between the classes, the genetic idea as regards sexual matters was (as still with savages in questions of food) to favour the seniors and defend their rights in defining each one's status. But actually, with the decay of incest, it would become what it is as among lower races, where nothing is more remarkable than the strict interdict upon any union between members of different generations.[324]

It is evident that hence complications might arise perplexing to the savage mind. For instance, we may expect to find cases where the niece is an adult, whilst the aunt is still an infant, and yet marriage between the former and the son of the latter is obligatory, as they are cousins of the same generation. Here, probably, we have a clue to one of the most bizarre facts in anthropology, where the universal rule as to sexual connection between generations seems to

---

[324] The most distinctive feature to-day in the inter-relations of generations is a most strict ordinance to celibacy between members of different generations.

be wantonly disobeyed, although in reality the reverse may be seen to be the case on examination. It is recorded of the Keddies of Southern India that a very singular custom exists among them, a young woman of sixteen or twenty years of age may be married to a boy of five or six years. She, however, lives with some other adult male (perhaps a maternal uncle or cousin), but is also allowed to form a connection with the father's relatives, occasionally it may be the boy's father himself, i.e. the woman's father-in-law! Should there be children from these liaisons, they are fathered on the boy husband. When the boy grows, the wife is either old or past child-bearing, when he in turn takes some other boy's wife in a manner precisely similar to that in his own case and procreates children for the boy husband.

By the classificatory system, as each in fact is a member of the same generation, they are born husbands and wives. The enforced virginity of the wife, implied under such conditions, entailed a celibacy incompatible with all lower ideas. It is easy to imagine the compromise between his conscience and his desires which a savage would make in such a case when favoured (or forced) by circumstances of environment, for it is unknown elsewhere. The infant nephew goes through the ceremony of marriage, which, by a fiction, being thus legally consummated, the wife is left free to follow her desires. These, however, are by no means allowed to run riot. They are regulated in a fashion of which, although the peculiarity is noted by the authors of the extract, the full significance can only be appreciated in connection with our hypothesis. She formed indeed connections outside of her husband, but solely with those of the legally eligible totem. As I believe the Keddies have male descent, these would be sons of the father's sister, or sons of the mother's brother, or again with the latter himself, who was her father-in-law, whereas union with the sons of the father's brothers, or of the mother's sisters, as being of the same totem, would not take place—and this we find to be the actual fact, as evidence proves.

But still other complications will be found to arise as the effect of the original concept of the classificatory system when brought face to face with new and advanced social order, which will have closer relation to our present argument. The distinctive feature in the economy of the primitive group in its relation to all other groups was mutual hostility. The instinctive distrust of strangers would be accentuated by the habitual hostile capture of females, for such groups, except in the case of the incest between father and daughter, were yet purely exogamous. But such mutual hostility implies isolation of each community. Thus all law evolved, as we have said,

226

would be purely with a view to regulation of the internal economy of a single consanguine group alone. Now in such a group, the division into generations of old male, male, and young male implied (although not as yet understood as between generations) the relationship of parents and children. Each generation is either child or parent to the other. As marriage is communal,[325] all the fathers in one generation are fathers to all the children in the next indiscriminately, and conversely these children recognise as fathers all the males of the senior generation. It follows that the relationship of all the members of a generation is purely fraternal, all are brothers and sisters to each other, and in this consanguine family they were really either actually so, or at least half brothers and half sisters.

Between these the primal law of celibacy between brother and sister as such embraced the whole generation. Now as long as the family was thus simply constituted, no friction would arise. The brothers, in common, captured and married in common some outside female,[326] and their children constituted solely the next generation. The sisters were either stolen or emigrated to other groups; but we have seen that a moment would come when this process ceased to be universal. The sister came to remain in her own group, and she was joined by some outside suitor; with the advent of their children, who are cousins to the others, would arise dire perplexities, in view of the old law.

We may now begin to see more distinctly, in the fact of the presence of the cousins, the resolution of the problem as to how a sister's son came to be also a brother's, and we will find that Mr. Morgan was not the first to be baffled by the problem. It was too intricate for primitive man at any rate. When first presented to him, we may surmise that he, in fact, refused to recognise it as a problem at all. Since the beginning of things in the group, as constituted by all tradition, the children of one generation were children of another simply, and nothing more. That as a result of the presence of the outside male, some intricate process of scission had occurred, and things were not as they seemed, was an idea far too abstract to be readily seized. All in a generation had been ever, to early man, brother and sister, and brother and sister they should continue.

We have seen in a past chapter that it was actually to the interest of senior male group-members, while incest reigned, that this condition of things should endure. It put at their sole disposal

---

[325] How can marriage be communal, granting Mr. Atkinson's views about sexual jealousy?—A. L.
[326] Where is sexual jealousy?—A. L.

227

the daughters of their brothers-in-law, and in the primal law placed a ban on sexual intercourse between all the younger male and female members, as constituting them brothers and sisters. As a factor in this case, however, the effect of incest was more or less temporary. The real agent in the tardy or non-recognition of the cousinship thus created, was the conservative force of old habit and tradition. We must remember that, in so early a group, personal descent as such was in no way recognised. Mere local contiguity alone constituted the sense of relationship, exogamy for instance took the form of local exogamy, for as all within a locality were (locally) relations, so all outside were, as strangers, free in marriage. While then so strong a sense of the value of contiguity continued, and was in practice, the evolution of an idea of non-relationship of two individuals with a common habitat would be too complex. Again, a recognition in fact implies a vast modification of the whole organisation of the group, which thus contains in cousins the elements of marriage within itself. But this is the latest and highest type of group and constitutes the tribe. We can understand that such a step was not taken at once by early man. Even when recognised we know that definition lags behind the event.

Thus in such a case as cited, and at the stage we are studying, if we find two cousins in presence, who are yet unrecognised as cousins, then, if nomenclature has taken place, we should find exactly the terms employed in the Malayan table which misled Mr. Morgan. A sister's son would be termed the brother's son, simply because the individual was as yet ignored, although existent, as a cousin, as members of the same generation they were brother and sister. Classes by generations alone were recognised.

Now as regards the validity of our assumption that relativity in age served as a means to determine privilege as to wedlock, proof can be furnished by certain nomenclatory features, as between members of a class or generation, to be found in the Malayan table in Ancient Society and elsewhere. This will afford, incidentally, strong negative proof of our theory as to non-union between brother and sister. It will also incidentally furnish the strongest negative evidence that, so far from brother and sister living in incest, as Morgan holds, brother and sister were regarded as quite apart in the sense of any sexual relation between them. It will be seen that there is a profound distinction made in address between inter-marriageable people and those between whom celibacy is enjoined.

Both Mr. Morgan and Mr. McLennan have drawn attention to the peculiarities in the terms of address as between 'brothers' and as between 'sisters.' It is curious that the full significance of the phenomena therein presented escaped two such keen intellects. We find here that terms of address as between persons of the same sex and of the same generation, and ergo brothers or sisters, present the very remarkable features that

(1) 'The age of the person spoken to compared with that of the speaker plays a very important part in the matter of denomination.'

(2) 'Such names refer not to the absolute age of the person addressed.'

(3) 'The relationships of brother and sister are conceived in the twofold form of elder and younger, and not in the abstract, and there are special terms for each among the Seneca Iroquois.'

(4) 'There is no name for brother and sister (Malayan system). On the other hand, there are a variety of names for use in salutations between "brother" and "sister" according to the age and sex of the person speaking in relation to the age and sex of the person addressed.'

(5) Among the Eskimo the form of the terms of relationship appears to depend, in some cases, more on the sex of the speaker than on that of the person to whom the term refers.

(6) In Eastern Central Africa, if a man has a brother and a sister, he is called one thing by the brother, but quite a different thing by the sister.

We will now illustrate the idea more completely by an extract of terms from the table of Hawaiian relationships in Ancient Society. An older or a younger brother is to a sister simply addressed or mentioned by the general term Kaiku nana, but to her, in address or mention of an older or a younger sister, they are respectively Kaik a'ana and Kaika-i-na. Again, an older or a younger sister is to a brother collectively Kaikuwaheena, but to him an elder or a younger brother is respectively Kaiknana and Kaikaina.

Now in view of our argument as regards the origin of these diversities in some sexual feelings, it is a most significant feature in these details of the terms of address that the expression of the relativity of age between the speakers is confined solely to the intercourse between members of the same sex. That a brother is the senior or the junior of Ego is carefully noted, but a sister is simply and vaguely a sister. Why? simply because whereas, by virtue of the primal law, no possible question whatever of mutual interest in sexual matters could possibly arise between a brother and a sister, on the other hand friction might hourly occur between brothers or between sisters. In fact, if our theory is correct, then, as questions of sexual privilege or precedence could cause jealousy between members of the same sex, distinctions would be necessary by definition of seniority when address took place between these, and in these cases alone, and this indeed we find to be the fact. As conclusive evidence we would cite the further important fact that these very same distinctions of senior and junior are used, inter se, between all those of the same totem [phratry] as now existing, but are never employed for their tribal cousins of the other totem [phratry]. And the reason is the same. The latter naturally do not marry (in groups formed of only two classes) [phratries] into the same totem [phratry] as the former, and thus there is no cause for jealousy or necessity of definition, whereas individuals of the same totem [phratry] are ipso facto group [potential] husbands of the same group [potential] wives, or are at least eligible in marriage with the same totem groups [phratries], and hence necessity for the exact definition by age of each one's rights.

Thus, as with other laws or institutions we have traced, we find a desire for distinction as regards rights in sexual union to be the genetic cause of the classificatory system both as regards the generation and its component members.

In all periods of transition which a process in change in progress implies, we expect to find cases where the conservative force of tradition from the past has delayed recognition of the too novel present, and we discover that circumstances have moved too rapidly for the intelligence of the times. If we keep this fact in view, we have thus seemed to find a natural explanation of the knotty point which was the cause of dispute between Mr. Morgan and Mr. McLennan,[327] and we may thus venture to say that each was both wrong and right in his views of the classificatory system in general. Each has mistaken a part for a whole, and they were ignorant that they were upholding two sides of the same question. Mr. Morgan was in error in assuming the system's too intimate connection with a determination of affinities in blood, in relation to which primarily, as we hope to have shown, it had really neither purpose nor aim, as also in his too hasty assumption of a consanguine family founded on brother and sister incest, based on a mere conjectural solution of a verbal detail, an assumption which he himself acknowledges had no other foundation.

Mr. McLennan was in error in maintaining that the classificatory system concerned terms of address alone. To quote his own words: 'What duties or rights are affected by the "relationships" comprised in the classificatory system? Absolutely none; they are barren of consequences, except indeed as comprising a code of courtesies and ceremonial addresses in social intercourse.' On the other hand, as we have tried to show, the system had precisely both intention and effect in regulation, as regards sexual feeling, which is

---

[327] Cf. Mr. Tylor, J. A. I. xviii. 3, 265, who expresses the same opinion.

the strongest passion in nature. And yet each disputant again was right in a degree, for, in later times, the classificatory distinctions really served as terms of address as regards the clan [tribe?], whilst again the primitive terms, which simply describe generations of persons in their relation to the group, were afterwards, by philological transmutation, to come to have a more definite meaning expressing the sense of the personal parent.

NOTE TO CHAPTER VIII

Group Marriage

The idea that 'group marriage' exists among the dusky natives of Australia, and that 'the group is the social unit as regards marriage' (as explained in the earlier part of this book), was introduced by Messrs. Howitt and Fison in their Kamilaroi and Kurnai (1880). Messrs. Spencer and Gillen, in their Natives of Central Australia (1899), support the views of Messrs. Fison and Howitt. 'Under certain modifications group marriage still exists as an actual custom, regulated by fixed and well recognised rules, amongst various Australian tribes' (p. 56). 'Individual marriage does not exist either in name or practice in the Urabunna tribe' (p. 63). Mr. Crawley argues, on the other hand, that individual marriage does exist among the Urabunna, 'though slightly modified' (Mystic Rose, p. 482). For each 'slight modification,' the husband's consent must be obtained. The system is regarded by Mr. Crawley, not as a survival of promiscuity, more or less modified, but as an 'abnormal development.' He believes in individual marriage, as, from the earliest known times, 'the regular type of union of man and woman.' 'One is struck by the high morality of primitive man' (pp. 483-484). What Mr. Atkinson meant by saying that 'marriage is communal,' I do not understand, as, on his theory, sexual jealousy must have prevented each man of a generation, in a group, from being equally the husband of each woman, not his sister. The young braves are supposed to bring in women captives from without, and to marry them 'communally,' Then what becomes of jealousy? They ought rather to have fought for their captive, on the principles of a golf tournament, the survivor and winner taking the bride. Mr. Atkinson never saw his work except in his manuscript, and might have made modifications on such points as this, where he seems to me to lose grasp of his idea, as in his theory of recognition of the children of 'the outside suitor,' he seems to bring male descent into action at a period when, as he asserts elsewhere, it was not yet recognised by customary law. On the Keddies (p. 287) I have no information, the author giving no reference. A. L.

APPENDICES

APPENDIX A

ORIGIN OF TOTEMISM

In the following village sobriquets from the south-western counties of England the people are styled 'eaters of' this or that.[328]

ENGLISH VILLAGE SOBRIQUETS

	Eaters	
Ashreigney	—	Dog-eaters
Morchard	—	'Burd'-eaters
Roseash	—	Whitpot-eaters
Sandford	—	Cheese-eaters
Moreton	—	Tatie-eaters
Paignton	—	Pudding-eaters
Churston	—	Liver-eaters

Compare with these the following sobriquets of Siouan old totem kins, counting descent in the male line.

SIOUAN NAMES OF GENTES

Eaters
Eat the scrapings of hides
Eat dried venison
Eat dung
Eat raw food
Among the Sioux we have also noted the sobriquets
Non-Eaters of
Deer

---

[328] Western Antiquary, vol. ix., pt. ii., p. 37, August, 1899.

Buffalos
Swans
Cranes
Blackbirds, etc.

These sobriquets of non-eaters are probably totemic: the Deer kin does not eat deer, nor does the Crane kin eat cranes, and so on. Totem kins are named from what they do not eat; many totem kins with male descent are nicknamed from what they do eat, or are alleged by their neighbours to eat.

GROUP SOBRIQUETS IN ORKNEY

In the following letter, which I owe to the kindness of Mr. Duncan Robertson, we read that, in Orkney and Shetland, local sobriquets are derived from what the people are alleged to eat. The tradition is, Mr. Robertson informs me, that each group is named after the edible plant or animal which it brought when engaged in building the Cathedral of Kirkwall.

Crantit House, St. Ola, Orkney,
Jan. 29, 1903.

Dear Mr. Lang,—My tyrannical doctor won't let me out yet, so that I have not been able to collect all the information I should like to get for you about the Orkney nicknames—or 'bye-names,' as they are called here.

Here follows the list as taken from Tudor's The Orkneys and Shetland, with alterations:

I. MAINLAND OR POMONA

Kirkwall	Starlings
St. Andrews	Skerry-scrapers
Deemess	Skate-rumples
Holm	Hobblers
Orphir	Yearnings
Firth	Oysters
Stromness	Bloody-puddings
Sandwick	Assie-pattles
Harray	Crabs (of old, sheep)
Birsay	Hoes = dog-fish
Evie	Cauld kail
Rendall	Sheep-thieves

II. SOUTH ISLES

Hoy	Hawks, Auks or Tammynories

233

Walls   Lyars (Manx Sheer-water)
Burray   Oily Bogies = the skin
   buoys used for herring nets

South Ronaldshay

a. Grimness   Gruties
b.   Hope   Scouties (Skuas)
c. Widewall   Witches
d.   Herston   Hogs
e. Sandwick   Birkies
f. South Parish   Teeacks (Lapwings)

III. NORTH ISLES

Gairsay   Buckies
Wyre   Whelks
Egilsay   Burstin Lumps
Rousay   Mares
Shapansey   Sheep
Stronsay Limpets
Sanday   Gruelly Belkies
North Ronaldshay   Selkies = Seals (also called
'Tangie Whessos') and
'Hides'
Eday   Scarfs = Shag or small
Cormorant
Westray Auks = the common Guillemot
Papa Westray   Doundies = spent Cod

These are all the names I know or can hear of in Orkney. I wrote to Mr. Moodie Heddle of Cletts on the subject, as I knew him to take a great and intelligent interest in all such topics; and I have a most interesting letter from him, of which I shall give you the gist. He says he has no doubt that the origin of the names is that which you suggest, though some of the names do not at first sight appear to bear this out. Kirkwall 'Starlings' are easily accounted for, assuming that there have always been as many starlings about Kirkwall as there are now. They may well have been eaten by the townsfolk. I have tried them, and their breasts are not at all bad.

Skerry-scrapers.—The allusion here is to men who live off shell fish, 'dilse,' etc. off the skerries. There are—or were—excellent oysters on the St. Andrews skerries. Mr. Heddle tells me he has heard a woman insulting a man by saying she supposed he would soon leave no limpets in a certain bay, meaning that he was too lazy to work for his living.

Skate-rumple is, of course, the skate's tail. Deemess is the nearest land to a famous piece of water for skate, known as 'the skate-hole.'

Holm 'Hobblers' I do not understand, but shall make some further inquiries. I have an idea it is a reference to some bird; Mr. Heddle thinks it has something to do with seals, but neither of us knows.

Yearnings are, of course, the dried stomachs of calves used for making cheese.

Oysters.—The bay of Firth was famous for its oysters till the beds were overfished and destroyed some thirty years ago.

Stromness 'Bloody-puddings'—Mr. Heddle suggests that the people bled their cattle twice or thrice a year and made 'puddings' of the blood. This, of course, was done in the Highlands at one time.

Assie-pattles.—Either those who lay in the ashes or, Mr. Heddle suggests, who ate cakes baked in the ashes. Before iron girdles came much into use cakes were baked on flat stones; and there is a hill, known as 'Baking-stone Hill,' where the people used to come for stones that would not split in the fire. The peats used in Sandwick have a very red ash, which colours all persons and things near it.

Harray 'Crabs.'—Harray is the only parish in Orkney which does not touch the sea, and the name is given in irony. The old 'tee-name' is said to have been 'sheep.' The story is told that some fishermen passing through Harray dropped a live crab. The men of Harray could not make it out at all, and sent for the oldest inhabitant, who was brought in a wheel-barrow. After gazing at the monster for a few moments he exclaimed: 'Boys, hid's a fiery draygon; tak' me hame!'

I suspect there is some other tee-name than 'sheep-thieves' for the Rendall people, but will try to find out and let you know.

Hoy 'Hawks'—Mr. Heddle, who was formerly proprietor s²xof Hoy, says he thinks 'auks' must have been the original word, as he believes 'tammy-nories' was the old name. 'Auk' is Orcadian for the common guillemot, and a 'tammy-norie' is a puffin. Both of these birds abound in Hoy.

Mr. Heddle also tells me that the old name of 'Lyars' for the people of Walls was to a great extent replaced by 'Cockles.' The 'lyars' were very common in Walls at one time, and were esteemed a great delicacy, but, Mr. Heddle tells me, were to a great extent killed out by the brown rat. He himself remembers men being bitten by rats when putting their hands into holes to look for young 'lyars.' Some three generations back enormous numbers of cockles were taken and eaten by the people of Walls, and they seem to have been called 'Cockles'—or, I presume, 'eaters of cockles'—in consequence.

Oily Bogies.—I hardly see how this can have been 'eaters of.' There might have been some old story to the effect that the Burray men stole and ate these buoys, but I never heard it.

South Ronaldshay has names for every district, which no other island but the Mainland has.

Gruties is, Mr. Heddle says, equivalent to 'Skerry-scrapers'—people who get their living from the 'grut' or refuse left in bights by the tide. ('Grut,' see Norse gröde = porridge or gruel.)

Scouties may be derived from the skua, though Mr. Heddle gives an unpresentable derivation. The word Birkies he did not know the meaning of, but asked two or three people, who all said the Sandwick people were so called 'because Sandwick was such a place for tangles coming ashore, and the people had such a habit of eating what they called "birken" tangles, i.e. the stout or lower ends of the large thick tangles.'

Burstin Lumps are a sort of preparation of oatmeal, once a very favourite dish in the Isles.

Rousay 'Mares'—There is an old tale of a Rousay man who, being a coward, killed his mare and hid inside her from his enemies. Mr. Heddle sends me an old rhyme on the subject:

> As the Rousay man said to his mare:
> 'I wish I were in thee, for fear o' the war;
> I wish I were in thee without any doubt,
> Were it Martinmas Day before I cam' out.'

The North Ronaldshay people did eat seals. Why Hides I do not know. Mr. Heddle here suggests it may have had to do with witchcraft, in which skins and especially

236

seals' flippers were much used. Within the last ten years a man pulled down and rebuilt his byre because of some 'ongoings with a selkie flipper.'

The names are very old and must be of Scandinavian origin.

Yours sincerely,
DUNCAN J. ROBERTSON

In addition to these names of 'eaters,' simple names of animals, we have shown in the text, are as commonly given to English villages as totemic names are given to the totem groups of savages.

ANCIENT HEBREW VILLAGE NAMES

In Robertson Smith's Kinship and Marriage in Early Arabia (p. 219) he says: 'I have argued that many place-names formed from the names of animals are also to be regarded as having been originally taken from the totem clans that inhabited them.' Now where totemism is a living institution I know no instance in which a locality is named from 'the totem clan that inhabits it.' The thing cannot be where female descent prevails, as many totems are then everywhere mixed in each local group. Where male descent prevails we do, indeed, get localities inhabited by groups mainly of the same totem name. But their tendency is to let the totem name merge in the territorial title, the name of the locality, as Messrs. Spencer and Gillen prove for the Arunta and Mr. Dorsey for the Sioux.

Having found no instance where a totemic group gives its totem name to the locality which it inhabits, I was struck by a remark of Dean Stanley in his Lectures on the History of the Jewish Church (p. 319, 1870). He there mentions the villages of Judah which were the scenes of some of Samson's adventures (Joshua xv. 32, 33; Judges i. 35). The villages of Lebaoth, Shaalbim, Zorah, respectively mean Lions, Jackals, and Hornets. Nobody eats any of these three animals, and they may be names of totem groups transferred to localities—though of this usage I know no example among savage totemists—or they may merely be old Hebrew village sobriquets, as in England and France.

On consulting the Encyclopædia Biblica, under 'Names' (vol. iii. 3308, 3316) we find that 'there can be no doubt that many place-names' in Palestine 'are identical with names of animals.' Those 'applied to towns' (we may read villages probably) are much more common in the south than in the north. We have Stags, Lions,

Leopards, Gazelles, Wild Asses, Foxes, Hyænas, Cows, Lizards, Hornets, Scorpions, Serpents, and so on. These may have been derived from old totem kins, though I think that theory improbable, or from the frequency of hornets or scorpions in this or that place, or the villagers' sobriquet may have become the village name. The last hypothesis has hitherto been overlooked. The frequency of animal and plant names in the Roman gentes, Fabii (Beans), Asinii (Asses), Caninii (Dogs), is an instance that readily occurs. These may be survivals of totemism or of less archaic sobriquets, while the totem names themselves, as we have argued, may have had their origin in sobriquets.

APPENDIX B

THE BA RONGA TERMS OF RELATIONSHIP

The hypothesis that the Australian terms of relationship, as they now exist, really denote status in customary law, may perhaps derive corroboration from the classificatory system as it appears among the Ba Ronga, near Delagoa Bay. Here the natives are rich, industrial, commercial, and polygamous to the full extent of their available capital. Polygamy, male kinship, and wife purchase, with elaborate laws of dowry and divorce, have modified and complicated the terms of relationship. They are described by an excellent authority, M. Henri Junod, a missionary.[329]

M. Junod has obviously never heard of the 'classificatory system' among other races, and his explanation of certain 'avoidances,' such as between the husband and his wife's brother, father, and mother, is probably incorrect (turning, as it does, on the laws of wife-price and divorce), though it appears now to be accepted by the Ba Ronga themselves. But what more concerns us is the nature of terms of relationship. These terms denote status in customary law, determined by sex and seniority. Among the Basuto, 'a man is otherwise related to his sister than to his brother; his children are related to their paternal otherwise than to their maternal uncles and aunts,' and to their cousins in the same style. Relative seniority, entailing relative social duties, is also expressed in the terms of relationship. The maternal aunt, senior to the

---

[329] Les Baronga, Attinger, Neufchâtel, 1898, pp. 82-87.

mother, is 'grandmother.' The children of my father's brother and of my mother's sister, are my 'brothers' or 'sisters;' the children of my maternal uncle and paternal aunt are not my 'brothers' and 'sisters.' The children of a man's inferior wives call the chief wife 'grandmother,' and the other wives, not their mother, 'maternal aunts.'[330] The son of my wife's sister is my 'son,' because I may succeed to her husband on his death, and his father calls me 'brother.' The maternal uncle is the mere butt of his nephew, the uncle's wives are the nephew's potential wives: he is one of the heirs to them. This kind of uncle (maternal) is not one of the tribal 'fathers' of the nephew, but the paternal uncle is, and is treated with the utmost respect. In brief, each name for a 'relationship' is a name carrying certain social duties or privileges, dependent on sex and seniority.

We have no such customary laws, and need no such names— the names are the result and expression of the Basuto customary laws. Had we such ideas of duty and privilege, then they would be expressed in our terms of relationship, which would be numerous. My maternal uncle would have a name denoting the man with whose wife I may flirt. The wife of my brother-in-law is the woman whom I must treat with the most distant respect. If I am a woman, my father's sister's husband (my 'uncle by marriage') is a man whose wife I may become, and so forth endlessly. Consequently there is a wealth of terms of relationship, just because of the peculiarities of Ba Ronga customary law.

---

[330] Op. cit. pp. 487-489.